Graphic Organizers for Math

by Daniel J. Barnekow
illustrated by Marcellus Hall

J. WESTON
WALCH
PUBLISHER
Portland, Maine

User's Guide
to
Walch Reproducible Books

As part of our general effort to provide educational materials that are as practical and economical as possible, we have designated this publication a "reproducible book." The designation means that purchase of the book includes purchase of the right to limited reproduction of all pages on which this symbol appears:

Here is the basic Walch policy: We grant to individual purchasers of this book the right to make sufficient copies of reproducible pages for use by all students of a single teacher. This permission is limited to a single teacher and does not apply to entire schools or school systems, so institutions purchasing the book should pass the permission on to a single teacher. Copying of the book or its parts for resale is prohibited.

Any questions regarding this policy or requests to purchase further reproduction rights should be addressed to:

Permissions Editor
J. Weston Walch, Publisher
321 Valley Street • P. O. Box 658
Portland, Maine 04104-0658

1 2 3 4 5 6 7 8 9 10

ISBN 0-8251-4257-1

J. Weston Walch, Publisher
P. O. Box 658 • Portland, Maine 04104-0658

Printed in the United States of America

Contents

Preface

This book was conceived as a kind of "Swiss Army knife" for mathematics teachers. The idea was to create a teaching resource that would be there when you needed it, help you accomplish a wide variety of tasks, and get you out of some tough teaching spots.

That was the idea. The result is this book. Between its covers, you'll find a quick, prepared, practical assignment for virtually every major topic you're likely to cover in your mathematics classes. What's more, each assignment is designed to teach, reinforce, and extend the key concepts of the mathematics curriculum. This was accomplished by combining a teaching technique that has proven pedagogical values with national mathematics curriculum standards.

You can draw on the student activity sheets in *Graphic Organizers for Math* nearly any school day. At your fingertips you'll have a ready-made, easy-to-use, pedagogically valuable lesson, on nearly any topic. The lessons will be popular with students, and are suitable for either in-class or homework assignments.

We wish you, and your students, the best of luck in your endeavors.

My appreciation is extended to Lisa French and to Holly Moirs. More thanks go to the mathematics teachers and, especially, to their students, without whose insights this book would not be possible.

—Daniel J. Barnekow

To the Teacher

A teacher once described graphic organizers as "sophisticated doodles." In a way, he was right. In fact, that may be the best way for you to think about graphic organizers and to present them to your students. You can find many jargon-laden articles and books that analyze graphic organizers, put forth new taxonomies, and labor to link them to psychological dynamics. These have their place, of course, but graphic organizers—essentially a simple teaching tool—have been over-analyzed, with the net effect of confusing rather than enlightening educators.

Graphic Organizers for Math is designed to cut through the jargon and give you a practical tool that you can put to use immediately. Spend a little time reading this introduction and thumbing through the graphic organizers, and you'll be ready to go.

Understanding Graphic Organizers

On a practical, classroom level, all you need to know about graphic organizers can be summed up in a few key points. As you use this book—or use graphic organizers in any educational context—keep these ideas in mind:

Graphic organizers are simply ways to organize information visually. This is a simple, straightforward, and accurate description of graphic organizers.

Graphic organizers are nearly always appropriate. Most people tend to think in visual terms, so graphic organizers are an appropriate way to organize information on a page.

Graphic organizers come in very many forms. Many attempts have been made to categorize graphic organizers and to identify them by type. You've probably heard of sequence chains, concept maps, webs, flowcharts, Venn diagrams, and so on. (You'll find many of these in this book.) Some of the best graphic organizers are combinations of these standard forms, and some are utterly unique.

Graphic organizers are never right or wrong, only better or worse. Assuming that the information presented, and its interrelationships, are correct, there are no "wrong" graphic organizers. However, some do a better job of presenting the same information than others.

Graphic organizers are not communicative, but conceptual. They are tools that help students acquire knowledge, and not a means of imparting knowledge. Obviously, graphic organizers are excellent communication tools, but in the classroom, you should focus on using them as a way for students to learn, not a way for students to express what they've learned to you.

Graphic organizers are concept-driven. The form of a graphic organizer should follow its function, not vice versa.

Content and Organization: Major Fields, Key Concepts, and Main Ideas

This book covers a wide range of mathematics topics, as a glance at the table of contents will show. The major sections of the book correspond to *major fields* taught in mathematics classes—for example, algebra. Within each major field, the graphic organizers emphasize *key concepts*—for example, the roles of variables. Moreover, each graphic organizer focuses on the *main ideas* of each key concept.

This organization enables you to use these graphic organizers throughout the year to help students achieve the principal learning objectives of your school's mathematics curriculum.

Correlation to National Standards

The content and organization of *Graphic Organizers for Math* was inspired and guided by a respected set of national standards: *Principles and Standards for School Mathematics*, issued by the National Council of Teachers of Mathematics. The major sections of the book correspond to these ten major content and process standards. Each graphic organizer in this book supports one or more specific NCTM standards for the secondary level.

How to Use the Graphic Organizers in This Book

Of course, you can use these graphic organizers any way you see fit—they are flexible tools. You can use them for basal instruction, review, and extension and enrichment. You can have students work in pairs or small groups to complete them. The organizers function equally well as homework and in-class assignments and are also excellent guides for classroom discussion.

A Lesson Cycle for Individual Graphic Organizers: Educators have learned that following a few simple steps will help their students get the most out of their graphic organizers. These steps, tailored to the content of this book, are presented below in, well, a graphic organizer!

1. Familiarize yourself with the graphic organizer and the teaching notes for it.

↓

2. Explain or remind students of what graphic organizers are, and why they're worthwhile. Emphasize the importance of organizing information.

↓

3. Present the specific graphic organizer. Point out its subject, its organizational framework, and the introduction, direction line, and questions.

↓

4. Model using the graphic organizer. Use examples. Consider giving students an example of what to include in each cell. If the graphic organizer calls for them to choose its topic, provide them with options.

↓

5. Assign the graphic organizer as an individual, paired, or group activity.

↓

6. Review students' work. Use the Key Questions in the Teaching Notes to generate classroom discussion or extend individual student learning.

How to Use the Graphic Organizers as a Set

Because *Graphic Organizers for Math* has fairly comprehensive topic coverage, keyed to national standards, you can use the organizers as curricular signposts, correlating them to the main points in your curriculum.

Also, because these graphic organizers emphasize main ideas in the key concepts of the major areas of secondary mathematics, they are excellent candidates for inclusion in student portfolios.

Teaching Notes

Teaching notes for each of the graphic organizers in this book are provided at the beginning of each major section in the book. The notes are organized in the following format:

[number]

[Title of Graphic Organizer]

<u>Objective</u>

Identifies the major learning objective of the graphic organizer.

<u>Key Questions</u>

Key questions that generate classroom discussion, guide students in achieving the learning objective, and extend teaching about the subject of the graphic organizer.

<u>Usage Notes</u>

Provide tips and techniques for using the graphic organizer in the most effective manner. Included when useful.

I. Number and Operations

NO-1: Types of Numbers

Teaching Notes

Objective: Students will identify many different types of numbers and note their use in the media and other aspects of everyday life.

Key Questions:

- What are the different types of numbers?
- When is each type of number used?
- Why is it important to be able to recognize numbers by type?

Usage Notes: This graphic organizer works well as a whole-class activity.

NO-2: Understanding Large Numbers

Objective: Students will identify the names and numerical and exponential expressions of very large numbers.

Key Questions:

- What is the numerical and exponential form of each number?
- How is the number of zeroes in a very large number related to its scientific notation?
- How can knowing word parts help you identify large numbers?

Usage Notes: This graphic organizer works well as a partners activity. Emphasize the importance of understanding scientific notation for application in other areas.

NO-3: Equivalencies of Fractions, Decimals, and Percents

Objective: Students will demonstrate an understanding of the equivalency of fractions, decimals, and percents and of their conversion from one form to another.

Key Questions:

- Does every fraction have an equivalent decimal and percentage form? Does every decimal have an equivalent fraction and percentage form? Does every percentage have an equivalent fraction and decimal form?
- What steps do you take to convert one form to another?
- Why is it important to be able to make these conversions?

Usage Notes: Coordinate the use of this graphic organizer with your regular classroom instruction.

NO-4: Types of Operations

Teaching Notes

Objective: Students will review the four basic arithmetic operations.

Key Questions:

- What are the four arithmetic operations?
- What is the purpose of each operation?
- What are some terms associated with each operation?

Usage Notes: Focus students' attention on the Taking Another Step activity, in which they concentrate on identifying problem spots in their application of operations. Work with students individually to address these areas.

NO-5: Properties of Operations

Objective: Students will define and provide the mathematical expression of the five laws of operations.

Key Questions:

- What is the definition of each law?
- What is the formulaic expression of each law?
- What device or devices can you use to memorize each law?

Usage Notes: Insist that students memorize these five laws. Pair students to quiz each other on them.

Name ______________________________ Date ____________ **Student Activity Sheet**

NO-1: Types of Numbers

Introduction: When you were little, you thought of **numbers** as simply "counters"—1, 2, 3, and so on. As your knowledge of math increased, you learned that numbers—like so many other things in life—are really of many different types. Knowing the different types of numbers is essential for learning more about math. Even more important, this knowledge is critical for your success after school in the "real world."

Directions: Think about numbers. Remember as many different *types* of numbers as you can (fractions, decimals, denominators, integers, and so on). Then complete the diagram below.

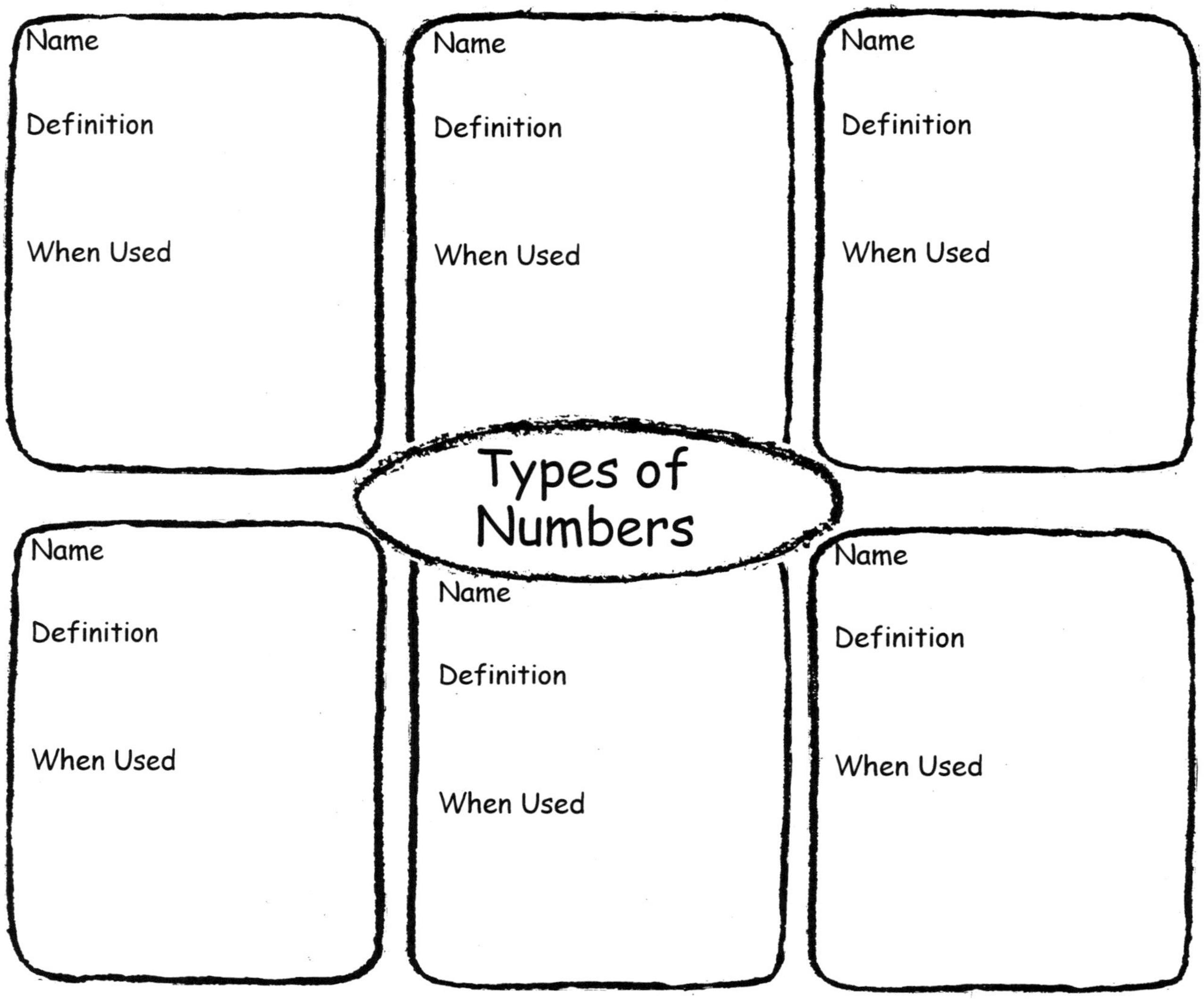

Taking Another Step: Look for examples of the use of different types of numbers in your daily life. You might find them on television shows, in magazines, and even in your conversations. Add boxes titled "Where I Saw One Last" to your diagram. Note where you last encountered each type of number.

Name ____ Date _____________ **Student Activity Sheet**

NO-2: Understanding Large Numbers

Introduction: Have you ever said something like this: "There were a gazillion of them!"? Do you know what a gazillion is? Do you know if *gazillion* is even a real number?

Words like *gazillion* are useful. They help you convey the idea that there are a lot—and we do mean *a lot*—of something. Scientists use such numbers all the time. They need them to talk about such things as the number of molecules in a substance, the number of cells in a body, or the number of light-years between stars. Of course, writing out very large numbers is tiresome. So, scientists write large numbers using a shorthand system called **scientific notation**.

Directions: Complete the table.

Very LARGE Numbers

Number Word	Number of Zeroes	Scientific Notation	Numeral
million			
billion			
trillion			
quadrillion			
quintillion			
sextillion			
septillion			
octillion			
nonillion			
decillion			
undecillion			
duodecillian			
tredecillion			
quattuordecillion			
quindecillion			
sexdecillion			
septendecillion			
octodecillion			
novemdecillion			
vigintillion			
googol			
centillion			
googolplex			

Taking Another Step: Use a dictionary to find the etymology (origin), of five of the number words in the table. Write a paragraph on the back of this sheet explaining why each of these numbers is named as it is.

Name ______________________________________ Date ____________ **Student Activity Sheet**

NO-3: Equivalencies of Fractions, Decimals, and Percents

Introduction: Which would you rather have: $^1/_{10}$ of a pie, .10 of a pie, or 10% of a pie? It really doesn't matter, since all of these statements are **equivalent**, or equal, to one another.

Fractions (like $^1/_{10}$), **decimals** (like .10), and **percents** (like 10%) are different ways of expressing quantities. In some situations, one system is more useful than another. For example, it is easier to offer a friend 1/8 of a pie than it is to offer .125 or 12.5% of it. Likewise, it is easier to say "drive .58 miles down the road" than it is to say "drive $^{29}/_{50}$ths of a mile down the road." As you can see, there are advantages and disadvantages to representing quantities as fractions, decimals, and percents. You need to know what each type of quantity means.

Directions: Think about when you have seen quantities expressed as fractions, decimals, and percents. Think about *why* each type of representation was used. Then complete the diagram below.

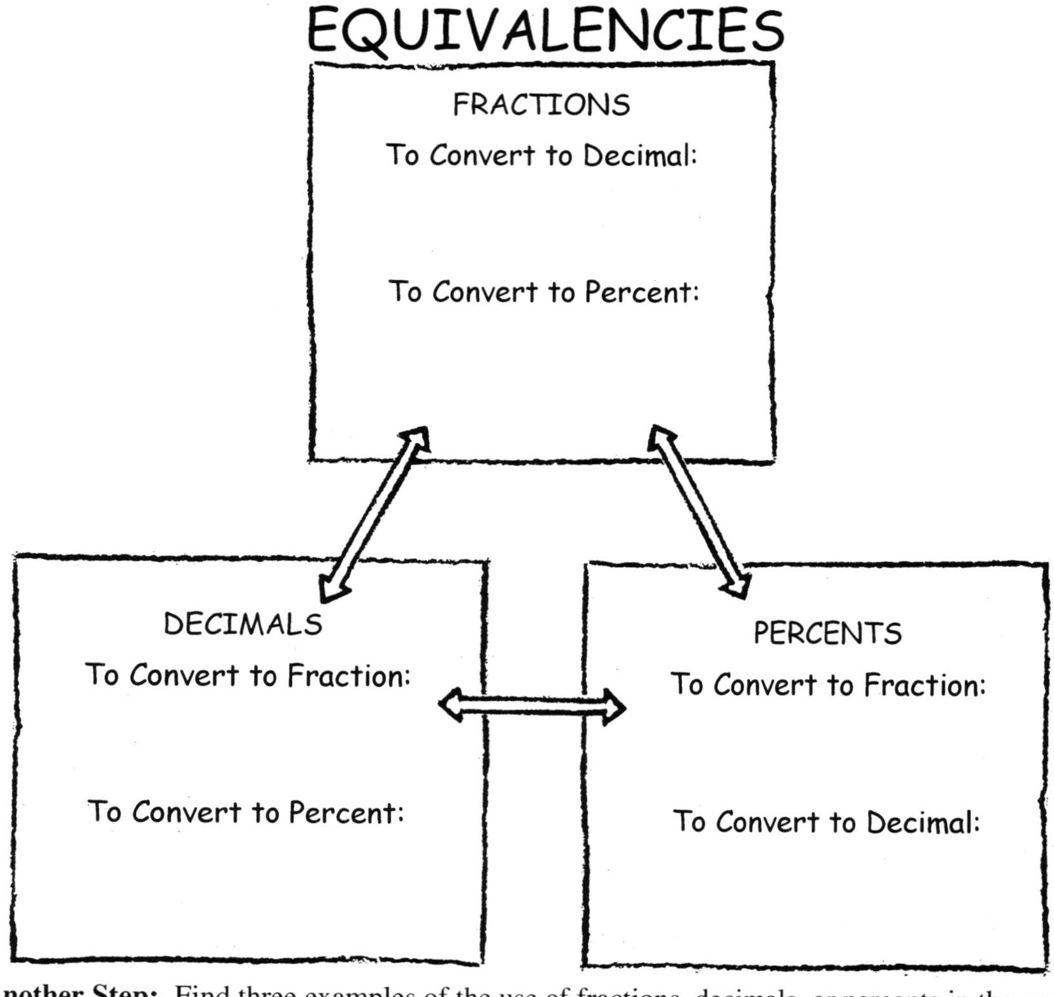

Taking Another Step: Find three examples of the use of fractions, decimals, or percents in the newspaper. Then convert each example into the other two types of quantity. Next, write an explanation that tells why the quantity was originally expressed the way it was.

Name________________________________ Date____________ **Student Activity Sheet**

NO-4: Types of Operations

Introduction: Arithmetic is the study of whole numbers, fractions, and decimals. It is also the study of their operations. **Operations** are the basic steps you take to solve problems in arithmetic. There are four basic operations in arithmetic: addition, subtraction, multiplication, and division.

Directions: Complete the diagram.

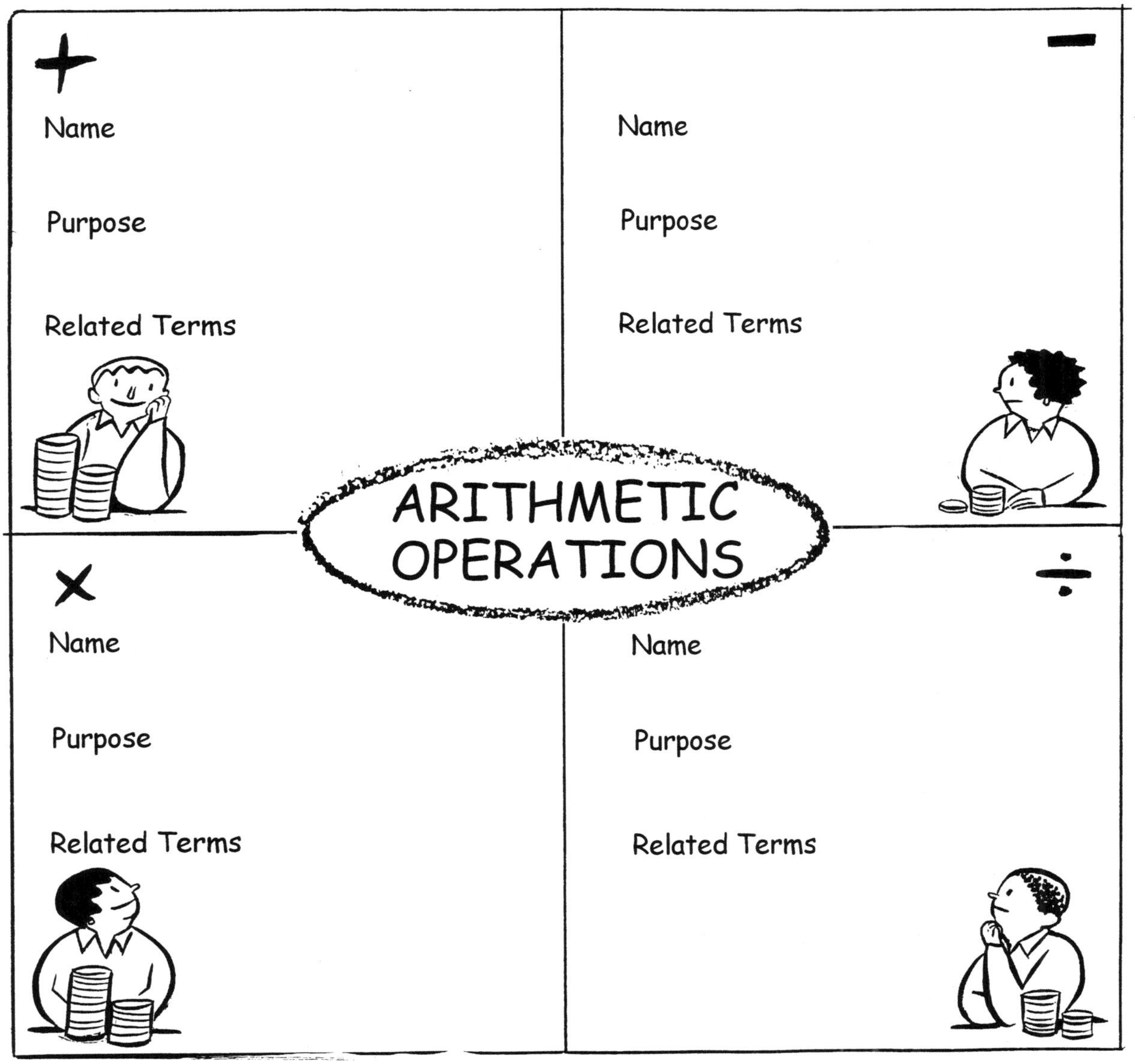

Taking Another Step: On the back of this sheet, write down an operation with which you sometimes have trouble. (For example, you might sometimes have trouble dividing fractions.) Then list two things you can do to become better at this operation.

Name ____ Date ____________ **Student Activity Sheet**

NO-5: Properties of Operations

Introduction: The famous mathematician Benjamin Peirce wrote, "Mathematics is the science which draws necessary conclusions." One reason the conclusions in math are "necessary," or unavoidable, is that certain rules must be followed. These math rules are so important that they are often called **laws**. The **properties of operations** are guided by five laws, which you should learn and memorize.

Directions: Complete the table below.

+ − The Five Laws of Operations × ÷		
Law	Definition	Expression
I. The Commutative Law of Addition		
II. The Associative Law of Addition		
III. The Commutative Law of Multiplication		
IV. The Associative Law of Multiplication		
V. The Distributive Law of Multiplication over Addition		

Taking Another Step: Think of ways to help yourself memorize the five laws of operations. For example, you might invent a silly sentence in which each word of the sentence begins with the same letter as one of the laws. Write your sentence or other memory device on the back of this sheet.

II. Algebra

A-1: The World Is Built on Algebra

Teaching Notes

Objective: Students will investigate algebra in a macro (larger) sense, defining it, its importance, and its use.

Key Questions:

- (Question forms of headings in graphic organizer.)
- What are some specific examples of the use of algebra in everyday life?
- What is your experience with algebra?

Usage Notes: This graphic organizer works well as a whole-class activity.

A-2: Terms Used in Algebra

Objective: Students will define and provide examples of terms used in algebra.

Key Questions:

- What does each term mean?
- What is an example that shows each term in action?
- How does the etymology of each term help you remember its meaning?

Usage Notes:

- Have students keep this graphic organizer in their notebooks as a quick-reference tool.
- Work with the class as a whole so that every student hears and takes notes on the etymology of each term.

A-3: Comparing Different Forms of Representation

Objective: Students will compare and contrast the advantages and disadvantages of three basic forms of representation used in algebra.

Key Questions:

- What are the advantages of each form of representation?
- What are the disadvantages of each form of representation?
- When is each form of representation the most useful? Why?

Usage Notes: Have students create each form of representation for a given problem to help them weigh the advantages and disadvantages.

A-4: Roles of Variables

Teaching Notes

Objective: Students will identify and investigate four roles that variables can play in algebra.

Key Questions:

- How can a variable play each role?
- What are examples of each role?

Usage Notes: Emphasize that identifying the role of a variable in a problem is one of the first steps students should take when addressing the problem.

A-5: Linear and Nonlinear Relationships

Objective: Students will compare and contrast linear and nonlinear relationships.

Key Questions:

- What is meant by a "linear relationship"?
- What is meant by a "nonlinear relationship"?
- Why is it important to understand the differences between the two?
- What are some examples of each?

Usage Notes: Ensure that students understand the proper use of a Venn diagram.

Name ______________________________ Date ____________ **Student Activity Sheet**

A-1: The World Is Built on Algebra

Introduction: If you threw a dart at a map of the world, there's a good chance you would hit a place where students are studying algebra, just like you. Why? Because algebra is studied all over the world. And that's because it is so very important. Algebra is used by scientists, engineers, business people, medical professionals, and others every single day. In fact, the world as we know it could not exist without algebra. This activity will help you see how this is so.

Directions: Use your textbook, talk to your teacher and classmates, and conduct other research to complete the diagram.

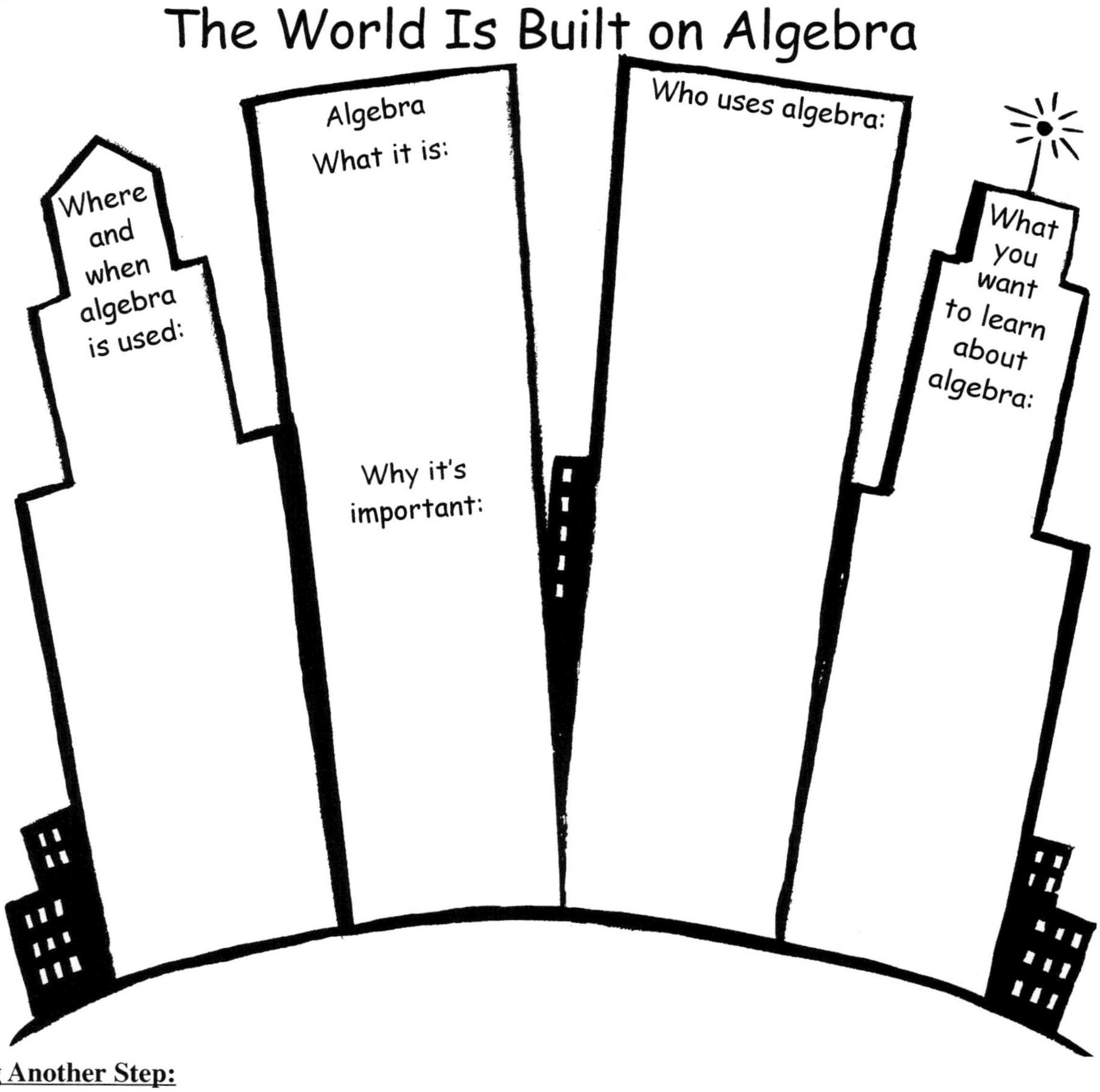

Taking Another Step:

Ask an adult how he or she uses algebra at work or at home. Write what you learn on the back of this sheet.

Name ____ Date _____________ **Student Activity Sheet**

A-2: Terms Used in Algebra

Introduction: Algebra is one of the chief branches of mathematics. Like the other branches (and, like most areas of knowledge outside of math), algebra has its own vocabulary. To understand algebra, you need to know this vocabulary. Luckily, the terms used in algebra make sense once you understand them. For example, a **variable** can vary in value, a **constant** doesn't change, and so on.

Directions: Fill in the "book." In the Notes column, write down something that will help you remember what each term means. Add any terms to the book that your teacher tells you to.

Terms Used in Algebra

Term	Definition	Example	Notes
Binomial	Algebraic expression consisting of two terms connected by a + or −	$a^2 + b^2$	is a certain kind of polynomial
Coefficient			
Constant			
Equation			
Exponent			
Expression			
Factors			
Monomial			
Polynomial			
Quadratic			
Root			
Term			
Variable			

Taking Another Step: Use a dictionary to find the etymology, or word origin, of three of the terms above. On the back of this sheet, explain why each of these terms is named what it is.

Name ______________________________________ Date ____________ **Student Activity Sheet**

A-3: Comparing Different Forms of Representation

Introduction: Which of the following is the *best* way to represent six of something?

six   6      5 + 1

The answer is that there *isn't* an answer. All of the examples convey the same information; they just do it in different ways. There are advantages and disadvantages to each. The same is true in algebra.

Directions: Think about when you have seen algebraic relationships expressed as tables, graphs, and equations. Think about *why* each type of representation was used. Then "weigh" the advantages and disadvantages of each by completing the diagram.

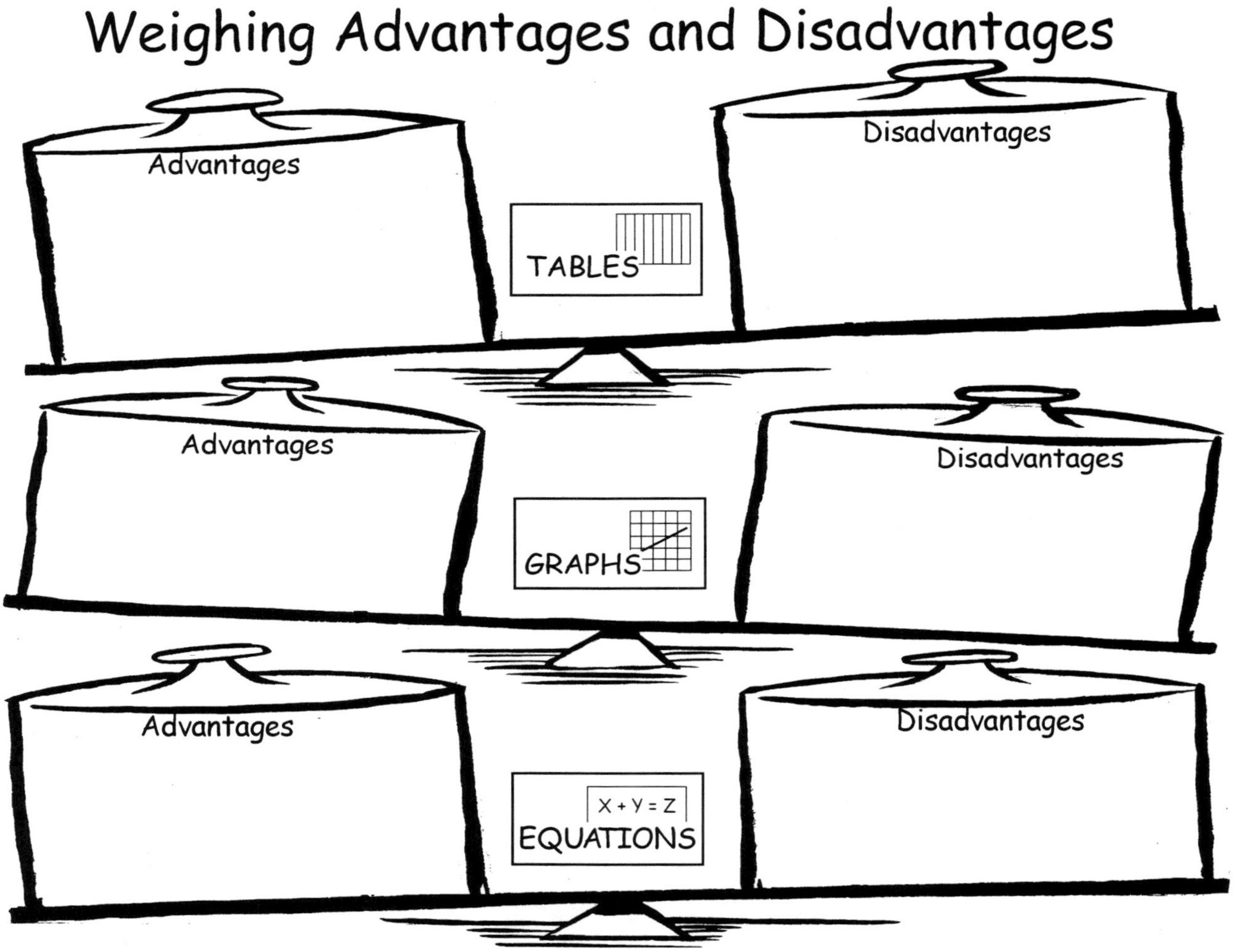

Taking Another Step: On the back of this sheet, draw three scales similar to the ones above. On one diagram, "weigh" the advantages of tables and the advantages of graphs against each other. On the second diagram, "weigh" the advantages of tables and the advantages of equations. On the third diagram, "weigh" the advantages of graphs and the advantages of equations. On the back of this sheet, write a paragraph that explains why you think each form of representation is best for certain situations.

Name ______________________________ Date ____________ **Student Activity Sheet**

A-4: Roles of Variables

Introduction: You might already know that, in algebra, a **variable** is a symbol that can be replaced by a numerical value. Usually, the symbol is a letter, like x. Often, the variable stands for an unknown number in an equation. But did you know that variables can do other things? Complete this activity to see just how versatile the variable is.

Directions: Complete the diagram.

The Versatile Variable		
The Variable Can . . .	**Example 1**	**Example 2**
Be a "**p**lace holder"		
Show an **a**rithmetic pattern		
Express a **f**ormula		
Show **c**ovariation		

Taking Another Step: Think of a way to help you memorize the four roles a variable can play. For example, you might invent a silly sentence in which each word of the sentence begins with the same letter as one of the roles. (Use the **bold** letters. For example, **P**eople **a**re **f**unny **c**reatures!") Write your sentence or other memory device on the back of this sheet.

Name____ Date____________ **Student Activity Sheet**

A-5: Linear and Nonlinear Relationships

Introduction: In many ways, algebra is about patterns. Two terms that describe algebraic patterns are **linear** and **nonlinear**. Linear means "in a line." A linear relationship can be represented by a solid line on an *x-y* coordinate graph.

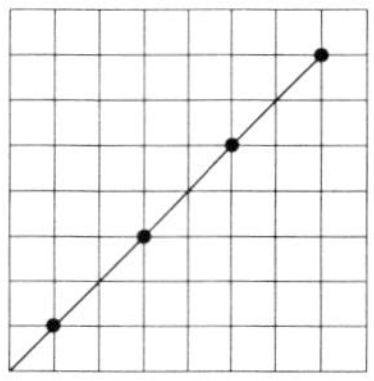

A nonlinear, or *discrete,* relationship cannot be represented by a continuous line.

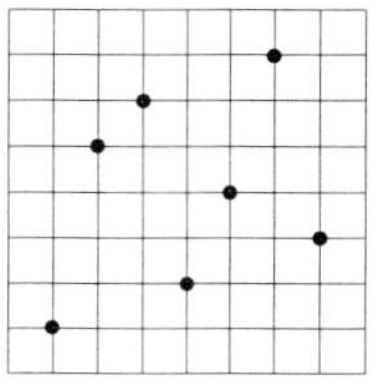

It's important to understand what each term means, and the difference between them.

Directions: As you learn about linear and nonlinear relationships, complete the diagram below with definitions, descriptions, examples, and other notes.

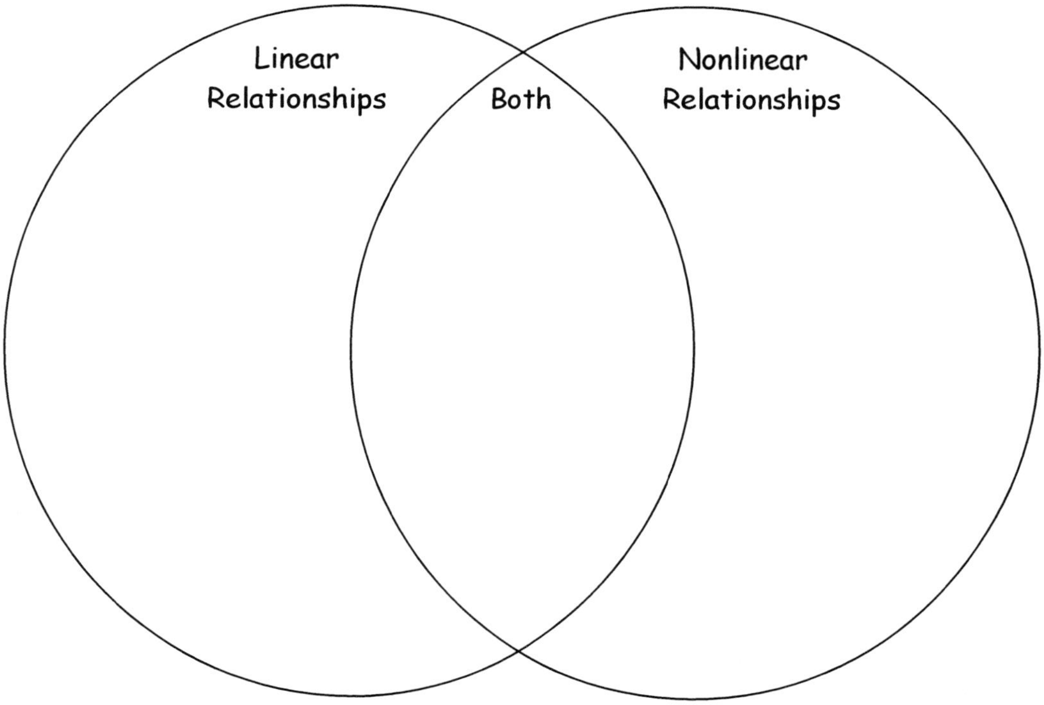

Taking Another Step: Which do you spend more time learning about: linear relationships or nonlinear relationships? On the back of this sheet, write a few sentences that explain why you think this is so.

III. Geometry

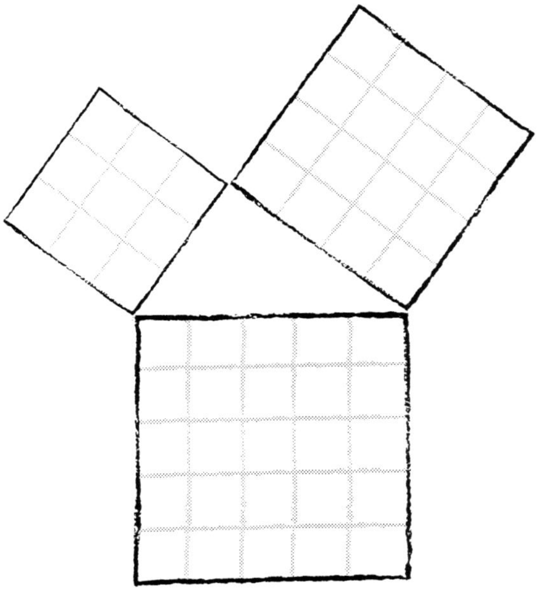

G-1: What Is Geometry?

Teaching Notes

Objective: Students will answer 12 fundamental questions about geometry, and then ask and answer 6 more.

Key Questions: (Questions on the graphic organizer.)

Usage Notes: Have students ask the questions they generate aloud, and use them as the basis for class discussions.

G-2: Terms Used in Geometry

Objective: Students will define and provide examples or illustrations of terms used in geometry.

Key Questions:

- What does each term mean?
- What is an example that shows each term in action?
- How does the etymology of each term help you remember its meaning?

Usage Notes:

- Have students keep this graphic organizer in their notebooks as a quick-reference tool.
- Work with the class as a whole so that every student hears and takes notes on the etymology of each term.

G-3: Characteristics of Two-Dimensional Shapes

Objective: Students will name a variety of two-dimensional shapes, identify their characteristics, and sketch them.

Key Questions:

- What shapes did you identify?
- What are the characteristics of each shape?

Usage Notes: Emphasize the "characteristics" cell for each shape. Students should describe qualities of angles, diagonals, sides, etc.

G-4: Characteristics of Three-Dimensional Shapes

Objective: Students will name a variety of three-dimensional shapes, identify their characteristics, and sketch them.

Key Questions:

- What shapes did you identify?
- What are the characteristics of each shape?

Usage Notes: Emphasize the "characteristics" cell for each shape: students should describe qualities of sides, their relationships, area formulas, etc.

G-5: The Pythagorean Relationship

Teaching Notes

Objective: Students will identify and describe the Pythagorean relationship.

Key Questions:

- What is the Pythagorean relationship?
- To what does it apply?
- What formula expresses it?
- Why might it be useful?

Usage Notes: For the Taking Another Step activity, explain to students that Egyptians used ropes stretched into triangles to form right angles to lay out their fields. The relationship of the lengths of the rope to the sides of the triangle ensured the creation of a right angle.

G-6: Geometry in Everyday Life

Objective: Students will identify and describe the use of geometry in their community.

Key Questions:

- What items did you see?
- How did geometry play a part in each one?
- Why is it important to learn about geometry?

Usage Notes: Emphasize the fundamental role geometry plays in the human-created world.

Name ____ Date ______________ **Student Activity Sheet**

G-1: What Is Geometry?

Introduction: Geometry is one of the chief branches of mathematics. It is concerned with the shape, size, and position of geometric figures, both two- and three-dimensional. You can quickly learn more about geometry by asking—and answering—a few questions.

Directions: Conduct research to answer questions 1 and 2 in each shape.

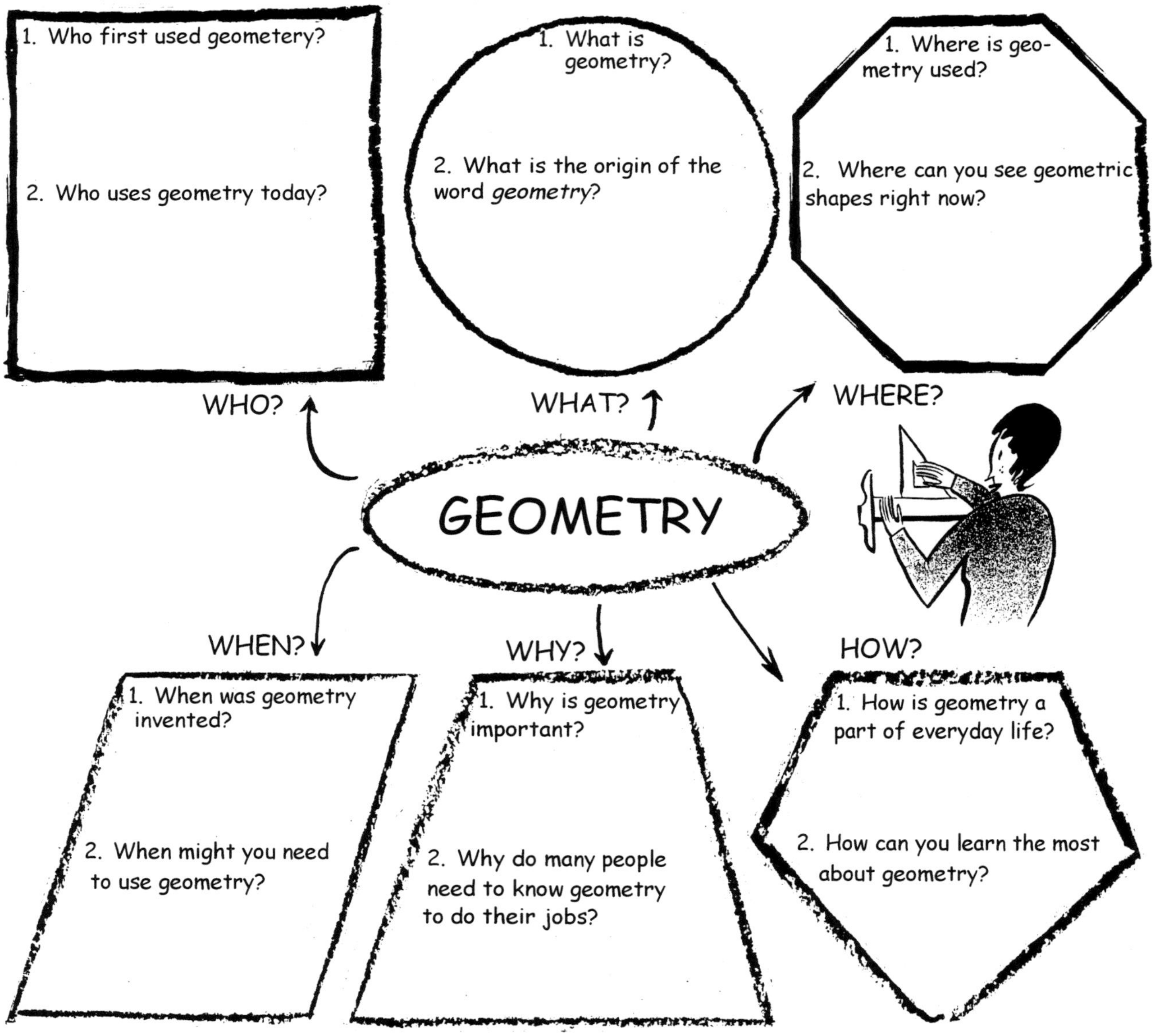

Taking Another Step: Ask six questions of your own and write them on the back of this sheet. Then trade your sheet with a partner and answer each other's questions.

Name ______________________________________ Date ____________ **Student Activity Sheet**

G-2: Terms Used in Geometry

Introduction: Geometry is one of the chief branches of mathematics. Like the other branches (and, like most areas of knowledge outside of math), geometry has its own vocabulary. To understand geometry, you need to know this vocabulary.

Directions: Fill in the "book" as you learn new geometric terms. In the Notes column, write down something that will help you remember what each term means.

Terms Used in Geometry

	Term	Definition	Example or Illustration	Notes
1.	plane figure	a form enclosing a space on a flat, two-dimensional surface	circles, triangles, squares	polygons are plane figures bounded by lines
2.				
3.				
4.				
5.				
6.				
7.				
8.				
9.				
10.				

Taking Another Step: Use a dictionary to find the etymology, or origin, of three of the terms above. On the back of this sheet, explain why each of these terms is named as it is.

Name ____ Date ____________ **Student Activity Sheet**

G-3: Characteristics of Two-Dimensional Shapes

Introduction: One of the basic types of shapes in geometry is the **two-dimensional shape**, or **flat shape**. There are many different types of flat shapes.

Directions: Complete and expand the diagram by adding boxes to the vertical line as you learn about different types of two-dimensional shapes.

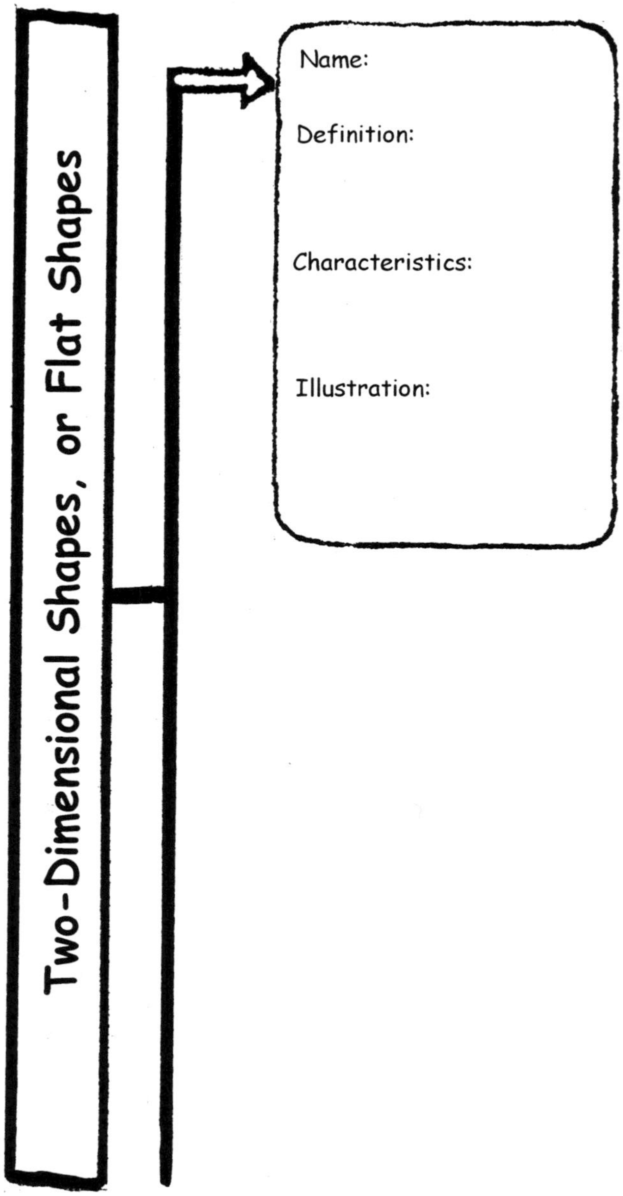

Taking Another Step: On the back of this sheet, write down the names of as many different two-dimensional shapes as you can find in your classroom or in your home.

Name ____ Date _____________ **Student Activity Sheet**

G-4: Characteristics of Three-Dimensional Shapes

Introduction: One of the basic types of shapes in geometry is the **three-dimensional shape**, or **solid**. There are many different types of solids.

Directions: Complete and expand the diagram by adding boxes to the vertical line as you learn about different types of three-dimensional shapes.

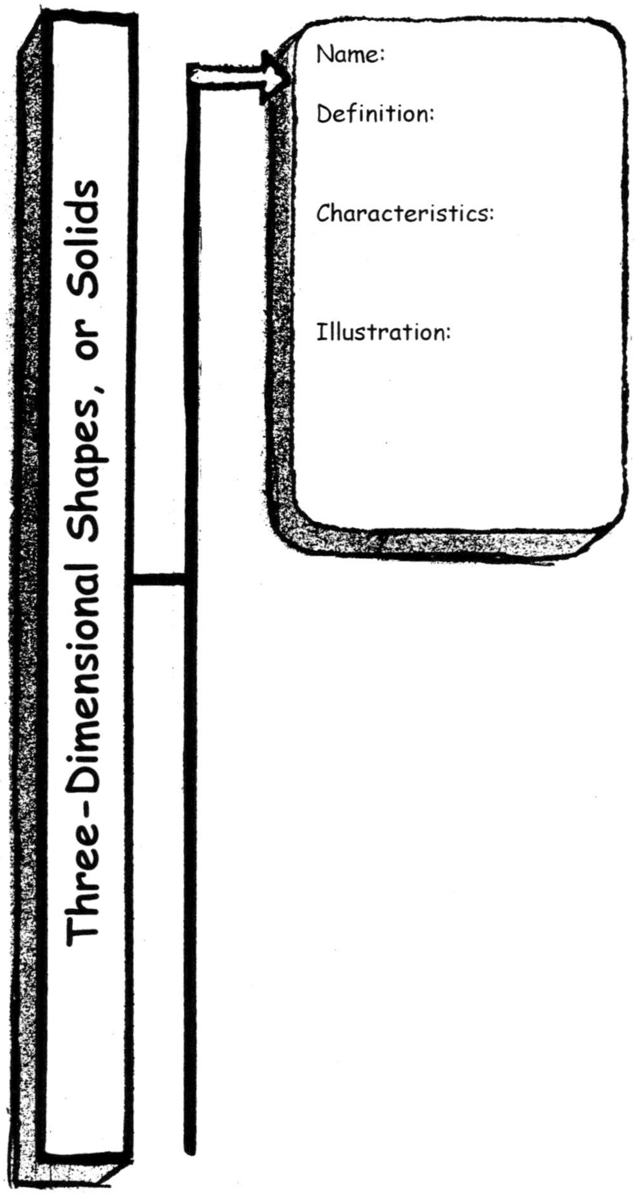

Taking Another Step: On the back of this sheet, write down the names of as many different three-dimensional shapes as you can find in your classroom or in your home.

Name ______________________________ Date ____________ **Student Activity Sheet**

G-5: The Pythagorean Relationship

Introduction: The Pythagorean relationship is named for Pythagoras, the Greek philosopher and mathematician who lived about 2,600 years ago. Pythagoras put together a **theorem** about the relationships in a right triangle. A theorem is a statement that has been proved true. Although this theorem is named for Pythagoras, the ideas it expresses were known by the ancient Egyptians.

Directions: As you learn about the Pythagorean theorem, complete the diagram.

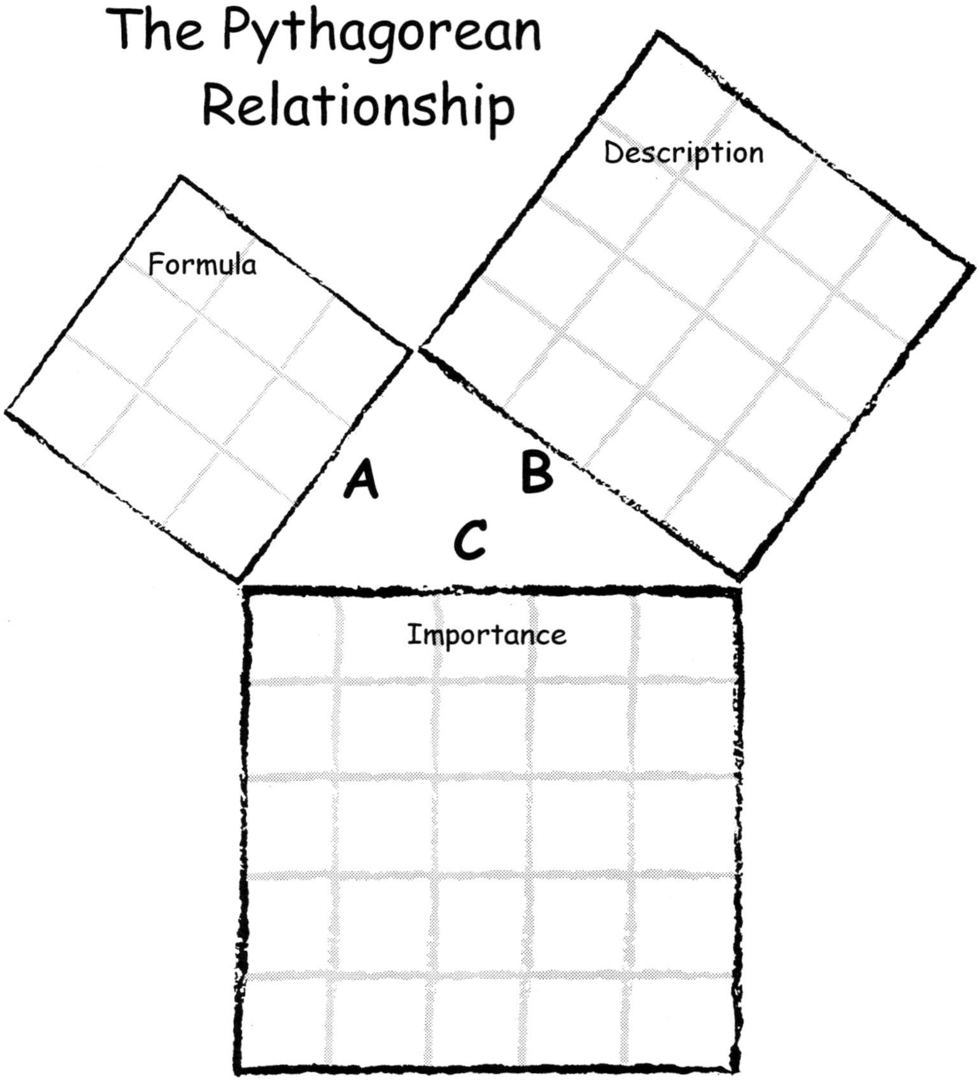

Taking Another Step: The ideas expressed in the Pythagorean theorem were first understood by the ancient Egyptians. They used these ideas to help them lay out farm fields. On the back of this sheet, write a paragraph explaining how you think the ideas behind the Pythagorean theorem would be useful for this purpose.

Name ______________________________ Date ____________ **Student Activity Sheet**

G-6: Geometry in Everyday Life

Introduction: Many students feel that math has little to do with "real life." In fact, most of the real lives that you and your classmates live are based, to a greater or lesser degree, on math. This is especially easy to see when you think about geometry. Your world is filled with lines and shapes put together in a way that gives the world an order. Your home, your clothes, the road you take to school, your school, and the many things you use every day are based on geometry.

Directions: Spend a day noticing how geometry is used in the things you use and see every day. For example, you might notice straight lines, right angles, and circles on a piece of notebook paper. You might see the triangular supports of a bridge in your community. As you notice the role of geometry in your life, fill in the log below.

Geometry in My World		
Item	**How Geometry Plays a Part**	**Notes**
1.		
2.		
3.		
4.		
5.		
6.		
7.		
8.		
9.		
10.		

Taking Another Step: Take a specific concept from geometry and, on the back of this sheet, write a paragraph that explains how it is used in the real world.

IV. Measurement

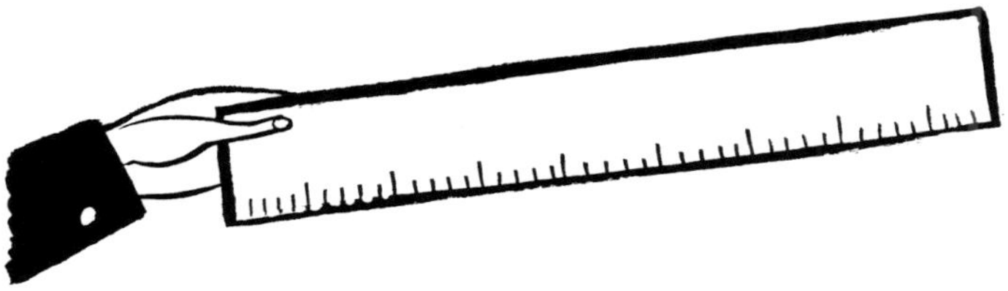

M-1: The Metric System

Teaching Notes

Objective: Students will explain the metric system as a decimal system, identify the four most common base units in the metric system (meters, liters, grams, square meters), and discuss the use of prefixes in the metric system.

Key Questions:

- (Question forms of the headings in the graphic organizer.)
- How is the metric system designed to make it easy to remember and use?

Usage Notes: Review the information on the graphic organizer with students with an eye toward student memorization of it.

M-2: Metric-English Conversions

Objective: Students will identify and apply the techniques for metric-English and English-metric conversions.

Key Questions:

- What do you multiply by to make each conversion?
- Why do you need to know how to make these conversions?
- What can you do to help yourself remember this information?

Usage Notes: Encourage student memorization of the information in the graphic organizer.

M-3: English and Metric Equivalencies

Objective: Students will identify the approximate metric equivalents of common English measurements.

Key Questions:

- Which are units of length? of area? of volume? of weight?
- What is the approximate metric equivalent of each one?
- How can this knowledge help you in everyday life?

Usage Notes: This graphic organizer works well as a partners activity or as a homework activity.

M-4: Perimeter, Area, and Volume

Objective: Students will define and compare and contrast the concepts of perimeter, area, and volume.

Key Questions:

- What is perimeter? area? volume?
- To what does each concept apply?
- How do you determine the perimeter, area, and volume of various shapes and parts of shapes?

Usage Notes: Encourage students to include formulas in their graphic organizers.

Name ______________________________ Date __________ **Student Activity Sheet**

M-1: The Metric System

Introduction: Every country in the world uses the metric system—except the United States! But that is slowly changing. More and more, the metric system is creeping into American life. This is because American government officials, and business people who deal with foreign businesses, must know the metric system. You need to know it, too.

Directions: "Go to the board" and complete the diagram.

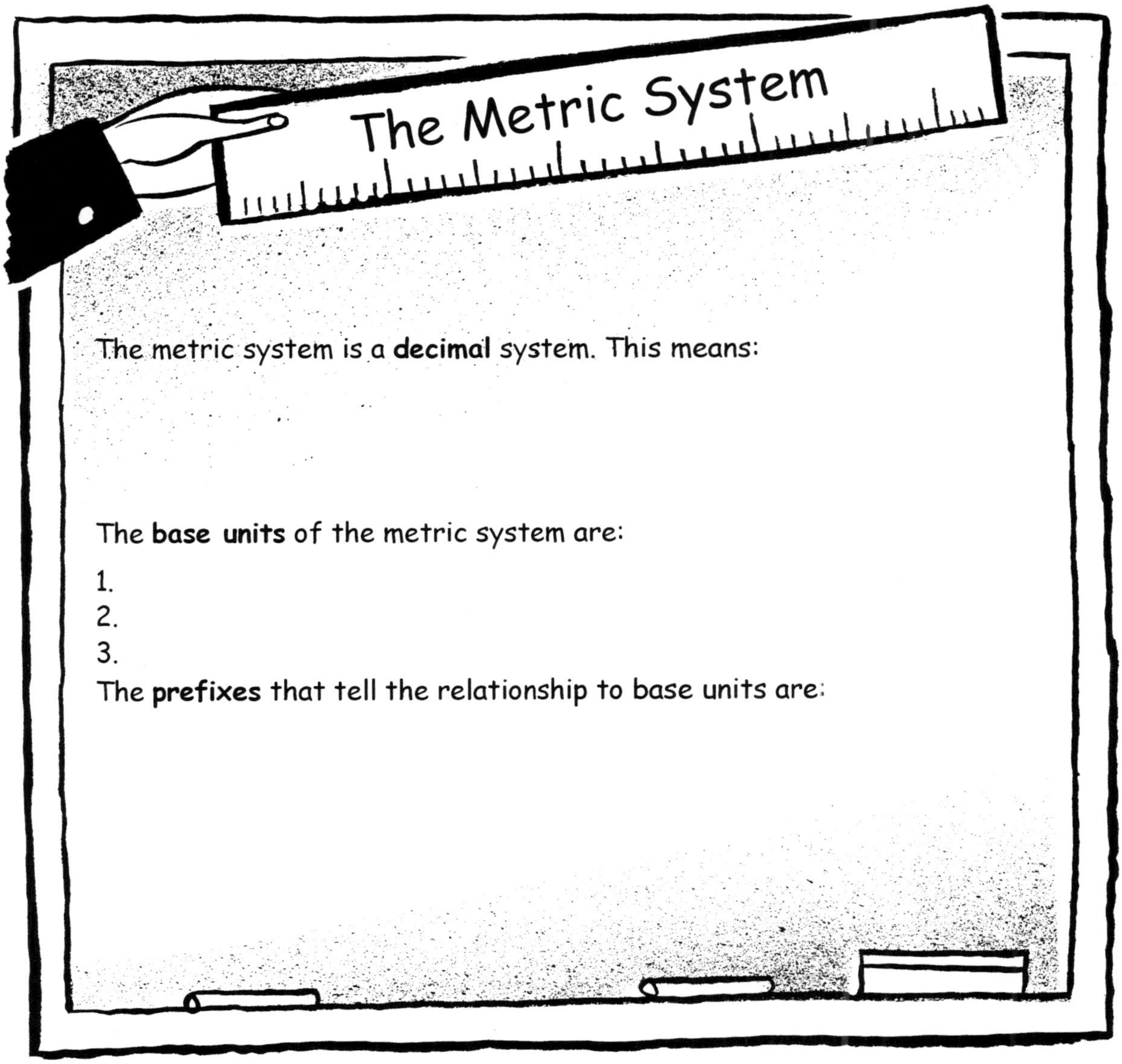

Taking Another Step: On the back of this sheet, combine several of the prefixes with several of the base unit names to create metric system terms. Check your terms in a dictionary or a textbook, and correct any that are wrong.

Name ____ Date _____________ **Student Activity Sheet**

M-2: Metric-English Conversions

Introduction: Many times you will need to convert an English (U.S.) measurement into a metric measurement, or vice versa. To make such a conversion, you need to perform some simple arithmetic.

Directions: Complete the metric conversion table.

Metric Conversion Table		
Length and Distance		
When You Know:	**Multiply by:**	**To Get:**
Area		
When You Know:	**Multiply by:**	**To Get:**
Volume		
When You Know:	**Multiply by:**	**To Get:**
Weight		
When You Know:	**Multiply by:**	**To Get:**

Taking Another Step: On the back of this sheet, list ten conversion problems: five English-to-metric, and five metric-to-English. Trade your list with a partner and complete each other's conversions. Check each other's answers.

Name ________________________________ Date ____________ **Student Activity Sheet**

M-3: English and Metric Equivalencies

Introduction: Would you rather walk 5 miles or 10 kilometers? Should you buy 2 liters of soda pop or six 12-ounce cans? Knowing the rough equivalent of English and metric measurements will help you in everyday life.

Directions: Complete the table.

English and Metric Equivalencies		
This . . .	. . . is about equal to. . .	. . . this.
one inch	≈	_______ centimeters
one foot	≈	_______ centimeters
one yard	≈	_______ meters
one mile	≈	_______ kilometers
one square inch	≈	_______ square centimeters
one square foot	≈	_______ square meters
one square mile	≈	_______ square kilometers
one acre	≈	_______ hectares
one ounce	≈	_______ milliliters
one pint	≈	_______ liters
one quart	≈	_______ liters
one gallon	≈	_______ liters
one ounce	≈	_______ grams
one pound	≈	_______ kilograms
one ton	≈	_______ metric tons

Taking Another Step: Go to the grocery store. Find a variety of items that are sold in English units. (For example, a 12-ounce can of soda pop, 3 square yards of wrapping paper, and so on.) On the back of this sheet, write down the name of the item and its English measurement. Then write down its approximate metric equivalent.

Name ______________________________________ Date ____________ **Student Activity Sheet**

M-4: Perimeter, Area, and Volume

Introduction: Perimeter, area, and volume are terms you will hear over and over again in your study of math, and in other subjects. They are also terms you will use throughout your life. Knowing exactly what they mean, how they are similar and different, and how to determine them is critical.

Directions: Complete the diagram. Write definitions, descriptions, formulas, and other information about perimeter, area, and volume in the spaces. If something you write applies to two of the measurements, or to all three, write it in an overlapping space.

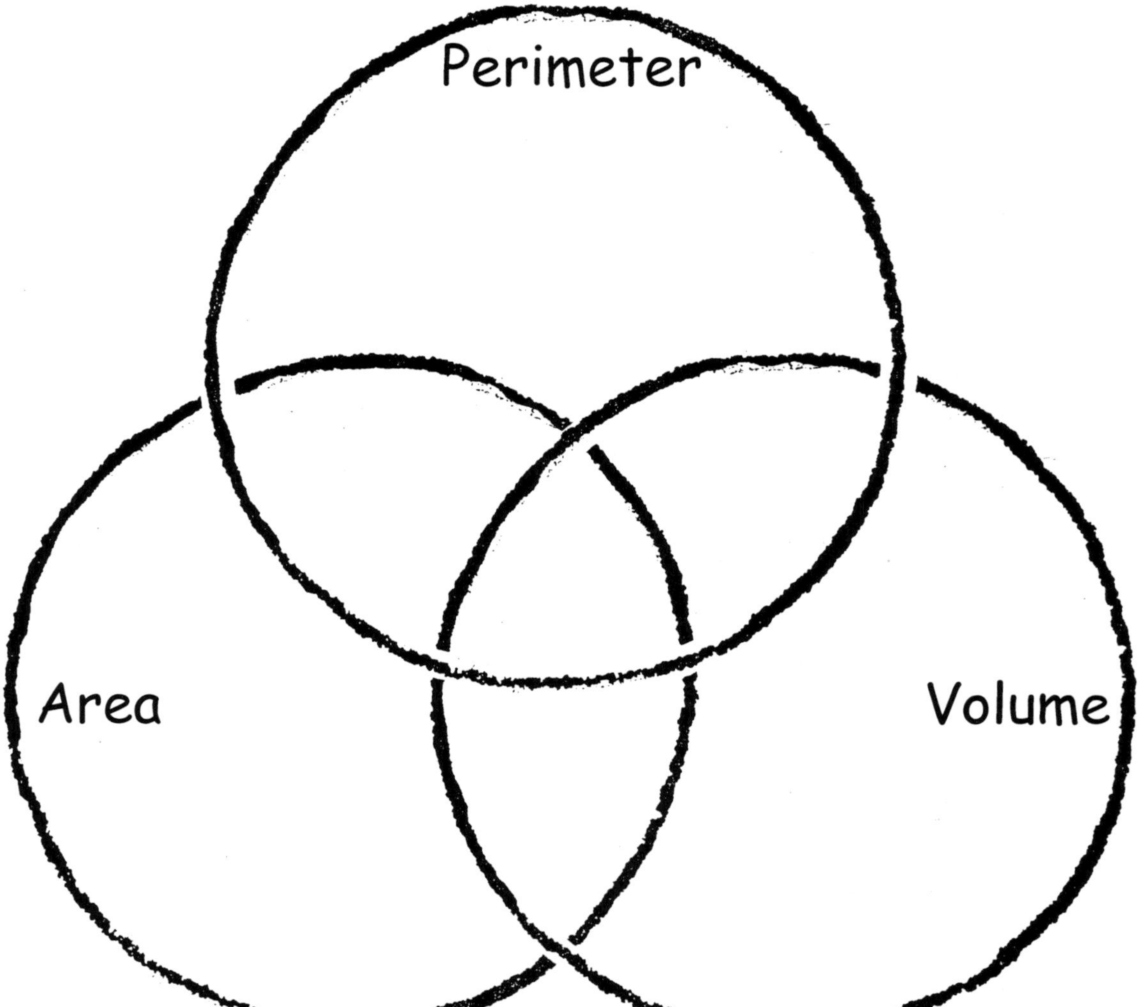

Taking Another Step: Identify one two-dimensional and one three-dimensional object in your classroom. For the two-dimensional object, determine its perimeter and area. For the three-dimensional object, determine its surface area and volume. Use the back of this sheet.

V. Data Analysis and Probability

DAP-1: Understanding Scatterplots

Teaching Notes

Objective: Students will analyze the construction of a scatterplot.

Key Questions:

- What is a scatterplot?
- What is the name of each part of a scatterplot?
- What is the purpose of each part of a scatterplot?

Usage Notes: Have students apply their macro knowledge of the parts and purpose of scatterplots to specific scatterplots in their textbooks.

DAP-2: Mean, Median, and Mode

Objective: Students will define, discuss, and apply their knowledge of mean, median, and mode.

Key Questions:

- What is a mean? a median? a mode?
- How are the three similar and different?
- How do you calculate mean? median? mode?

Usage Notes: Encourage students to use these terms explicitly where, in the past, they would have used the term "average."

DAP-3: What Is Probability?

Objective: Students will answer 12 fundamental questions about probability, and then ask and answer 6 more.

Key Questions: (Questions on the graphic organizer)

Usage Notes: Have students ask the questions they generate aloud, and use them as the basis for class discussions.

Name ______________________________ Date ____________ **Student Activity Sheet**

DAP-1: Understanding Scatterplots

Introduction: **Scatterplot** is a funny-sounding word that you will hear a lot. (Some people call scatterplots *scattergrams* or *scattergraphs*—they all mean the same thing.) The purpose of a scatterplot is to show where data points fall in relation to two variables. Scatterplots are fundamental to your understanding of many mathematical concepts, so it's important to be able to read them.

Directions: Label each part of the scatterplot below.

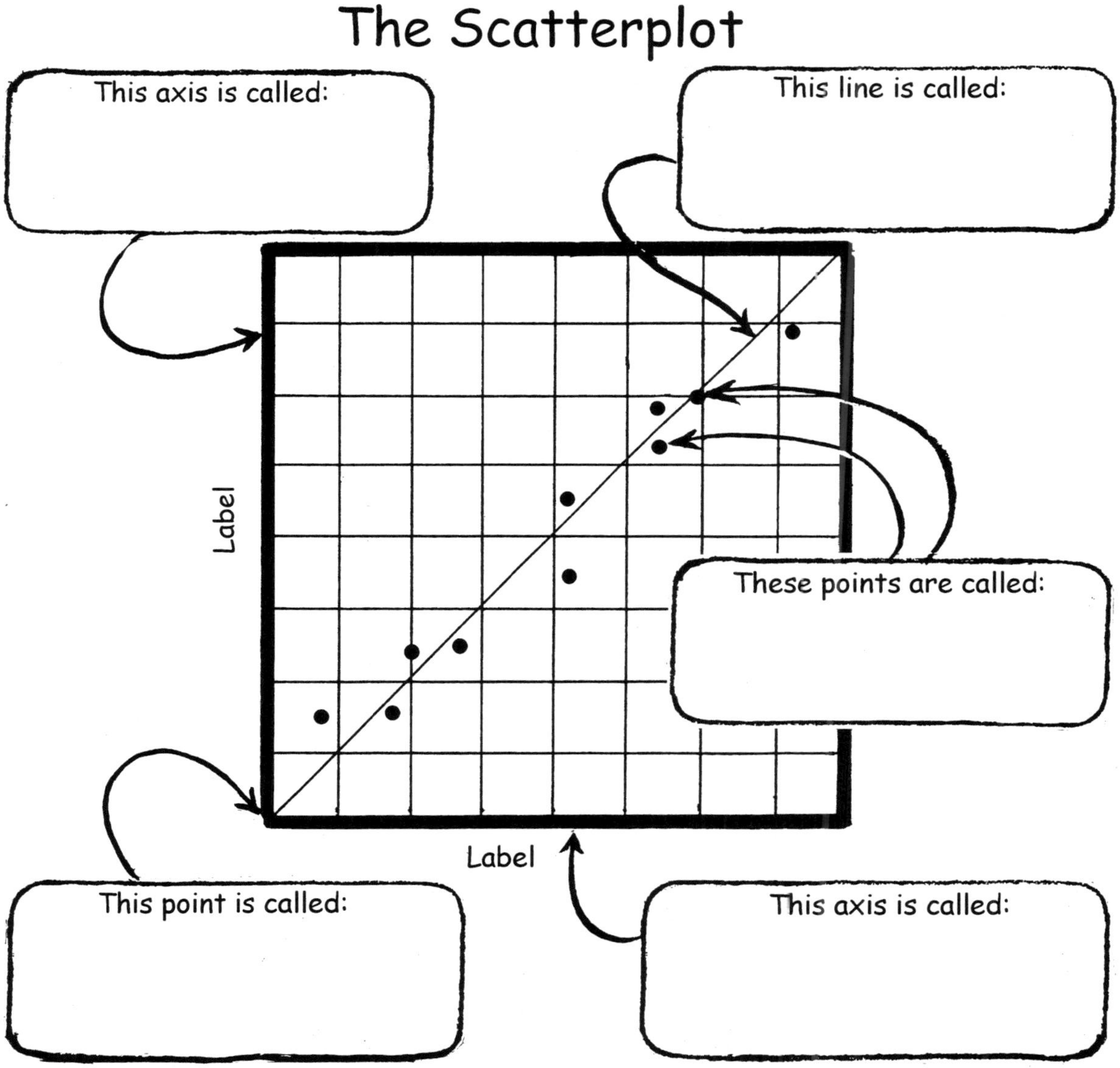

Taking Another Step: On the back of this sheet, construct a scatterplot that shows the relationship between the heights and widths of ten books.

Name ______________________________ Date __________ **Student Activity Sheet**

DAP-2: Mean, Median, and Mode

Introduction: You know the word *average*, but do you know that there are really three different kinds of averages? The three are the **mean**, **median**, and **mode**. People usually are thinking of *mean* when they use the word *average*. But median and mode are important, too, and can be just as useful. Make sure you understand how the three averages are similar and different.

Directions: Complete the diagram. Write definitions, descriptions, formulas, and other information about mean, median, and mode in the spaces. If something you write applies to two of the averages, or to all three, write in an overlapping space.

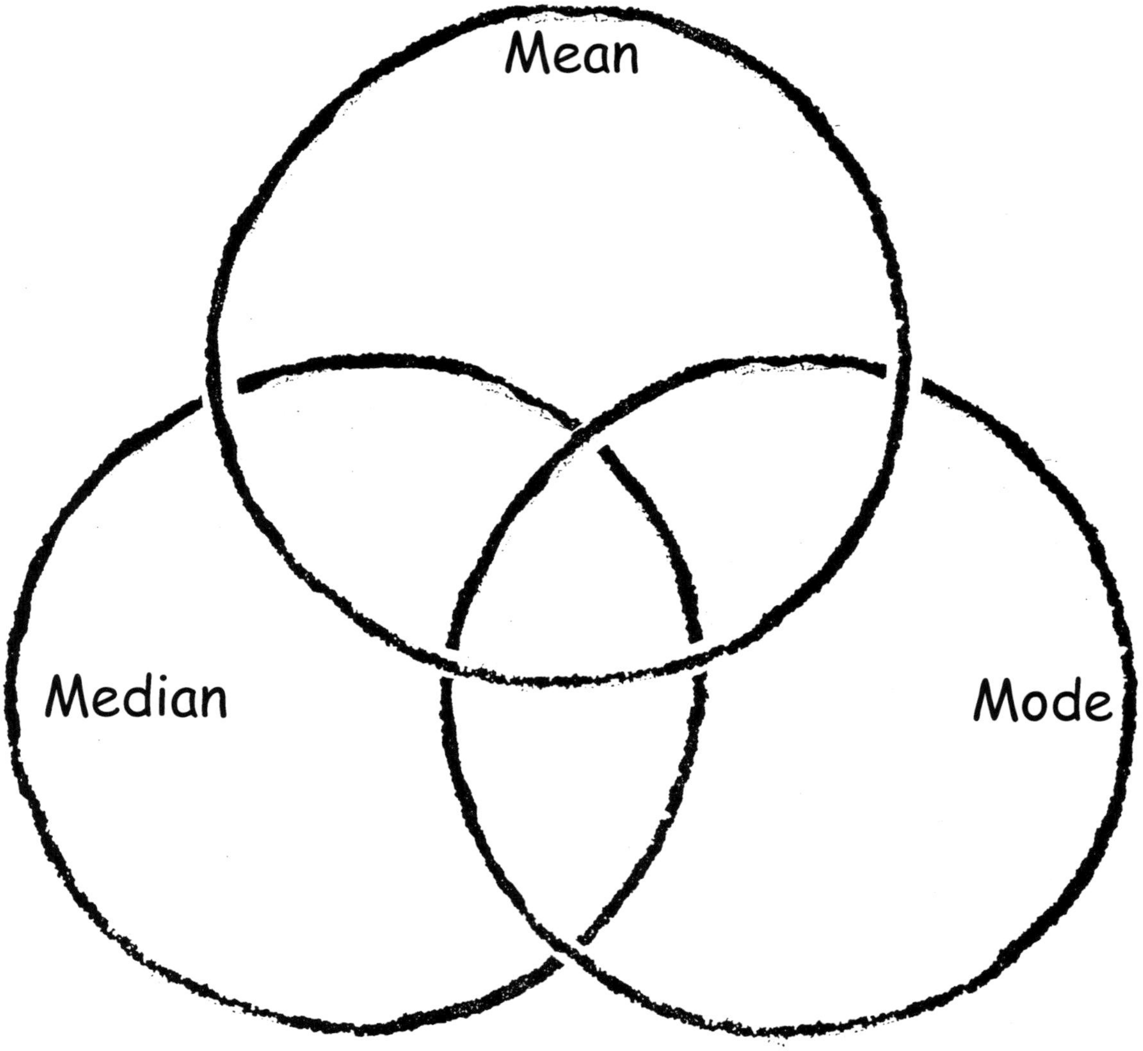

Taking Another Step: On the back of this sheet, list 13 numbers between 1 and 100. Trade your sheet with a partner. Now calculate the mean, median, and mode of each other's lists. Check your partner's work.

Name ____ Date _____________ **Student Activity Sheet**

DAP-3: What Is Probability?

Introduction: **Probability** is one of the branches of mathematics. It concerns the expression in numerical terms of the likelihood of events. Probability forms the foundation for statistics.

Directions: Conduct research to answer questions 1 and 2 on each die.

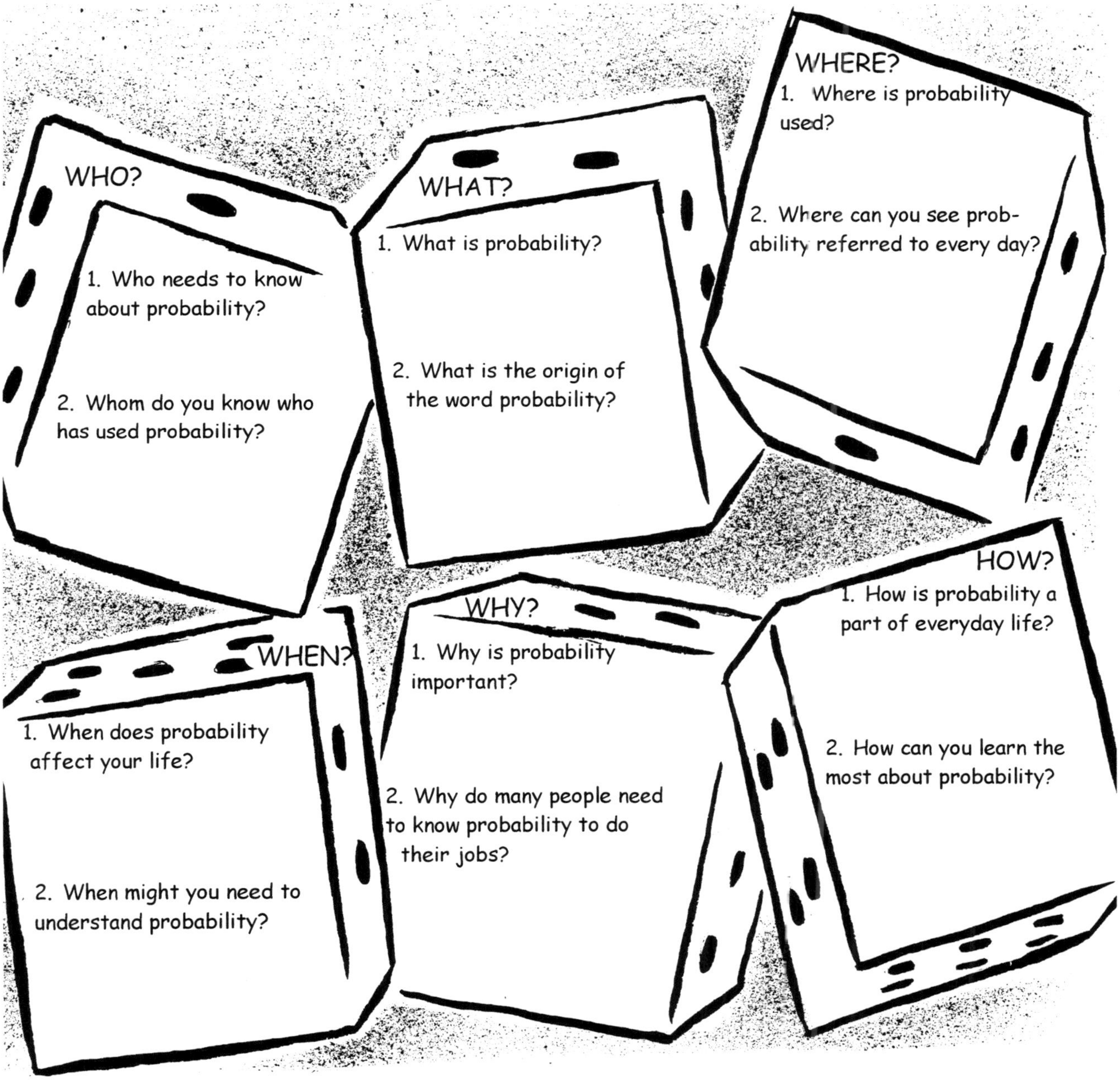

Taking Another Step: Ask six questions of your own and write them on the back of this sheet. Then trade your sheet with a partner and answer each other's questions.

VI. Problem Solving

PS-1: Problem-Solving Strategies

Teaching Notes

Objective: Students will identify and analyze four generic problem-solving strategies.

Key Questions:

- What problem-solving strategies did you identify?
- How does each one work?
- When have you used each one?

Usage Notes: Problem-solving strategies to consider include: looking for patterns; solving a similar, simpler problem; breaking the problem into smaller parts; working backwards; sketching a table or graph; and so on.

PS-2: K-W-L Chart

Objective: Students will apply a K-W-L chart to a mathematics activity or assignment.

Key Questions:

- (Questions on the graphic organizer)
- How can you use a K-W-L chart in the future?

Usage Notes: Encourage students to see the value of K-W-L charts, and to use K-W-L charts repeatedly in their academic careers.

PS-3: Problem-Solving Myths

Objective: Students will debunk five common myths about problem solving.

Key Questions:

- Why do you think each myth has developed?
- What truth should replace each myth?
- What steps can you take to concentrate on the truth and forget the myth?

Usage Notes: Consider assigning this graphic organizer very early in the school year.

PS-4: Myself as a Problem-Solver

Objective: Students will identify their strengths and weaknesses as problem solvers and determine ways to overcome their weaknesses.

Key Questions:

- What are your strengths as a problem solver?
- What are your weaknesses as a problem solver?
- What can you do to overcome each weakness?

Usage Notes: Be careful to respect students' privacy. Suggest specific, concrete ways for individual students to address each problem-solving weakness.

Name ____ Date ____________ **Student Activity Sheet**

PS-1: Problem-Solving Strategies

Introduction: How do you solve a math problem when no one has shown you how to do it? The answer is that you apply a problem-solving strategy that you already know. **Problem-solving strategies** are general ways to approach problems to find solutions. When you don't know the specific way to solve a problem, a general strategy can put you on the right path. It can also lead you to a solution.

Directions: Think about the different paths you have taken to solve problems in the past. Then complete the diagram by identifying and describing a general problem-solving strategy on each path.

Taking Another Step: On the back of this sheet, write about a time when you applied one of these problem-solving strategies successfully.

Name ______________________________ Date ____________ **Student Activity Sheet**

PS-2: K-W-L Chart

Introduction: K-W-L stands for What I ***K***now, What I ***W***ant to Know, and What I ***L***earned. Filling out a K-W-L chart before, during, and after a math assignment is a good way for you to keep track of your progress.

Directions: Complete the K-W-L chart.

Activity or Assignment: ______________________		
K **(Complete before)**	**W** **(Complete before)**	**L** **(Complete during and after)**
What do you already *know* about the activity or assignment?	**What do you *want* to learn about the activity or assignment?**	**What did you *learn* about the activity or assignment?**

Taking Another Step: On the back of this sheet, write a paragraph explaining how using a K-W-L chart did or did not help you with your assignment.

Name ______________________________________ Date ____________

Student Activity Sheet

PS-3: Problem-Solving Myths

Introduction: Math teachers know that, when it comes to solving math problems, many students are their own worst enemies. This is because many students hold mistaken beliefs about problem solving. These myths can make math more difficult than it needs to be. So take some time and replace these myths with the truth. You will be doing yourself a big favor!

Directions: Complete the table.

Problem-Solving Myths

The Myth	The Truth	What I can Do to Forget the Myth and Concentrate on the Truth
"All math problems can be solved quickly and directly, if you know how."		
"If you can't see right away how to solve a math probem, you won't be able to solve it."		
"There is one and only one right way to solve any math problem."		
"The best way to solve a math problem is to dive right in—sitting and thinking about it beforehand is a waste of time."		
"Either you have a gift for math or you don't. If math doesn't come easily to you, you will never succeed at it."		

Taking Another Step: Share your completed table with a classmate. Compare and contrast your entries in the last column.

Name ____ Date ____________ **Student Activity Sheet**

PS-4: Myself as a Problem Solver

Introduction: As a math student, you are a problem solver. And as a problem solver, you have both strengths and weaknesses. By identifying and thinking about them you can improve your math performance.

Directions: Complete the diagram.

My Strengths!

My Weaknesses

Ways to Overcome Each Weakness
and Make Them Strengths!

Taking Another Step: Arrange to discuss, privately, any problem-solving weaknesses you have with your teacher.

VII. Reasoning and Proof

RP-1: Deductive and Inductive Reasoning

Teaching Notes

Objective: Students will compare and contrast deductive and inductive reasoning.

Key Questions:

- What is deductive reasoning?
- What is inductive reasoning?
- How are the two types of reasoning similar? How are they different?
- What are some examples of each?

Usage Notes:

- Encourage students to apply their understanding of deductive and inductive reasoning to areas beyond the classroom.
- This graphic organizer works well as a whole-class activity.

RP-2: Mathematics Reasoning

Objective: Students identify and explain the basic steps in mathematics reasoning.

Key Questions:

- What does each step mean?
- What is an example of how you can apply each step?

Usage Notes: This is an advanced activity for students working alone. Consider reserving it for an enrichment exercise, or present it as a whole-class activity.

Name ____ Date ____________ **Student Activity Sheet**

RP-1: Deductive and Inductive Reasoning

Introduction: There are two basic types of reasoning: **deductive reasoning** and **inductive reasoning**. You will use both in mathematics. You will use both in other subjects. In fact, you will use both types of reasoning throughout your entire life. Therefore, understanding their differences is an absolutely vital part of your education.

Directions: Think about what you are learning about deductive and inductive reasoning. Complete the diagram. Under each name, write a definition and other information about that type of reasoning. Where the circles overlap, write information that applies to both types of reasoning.

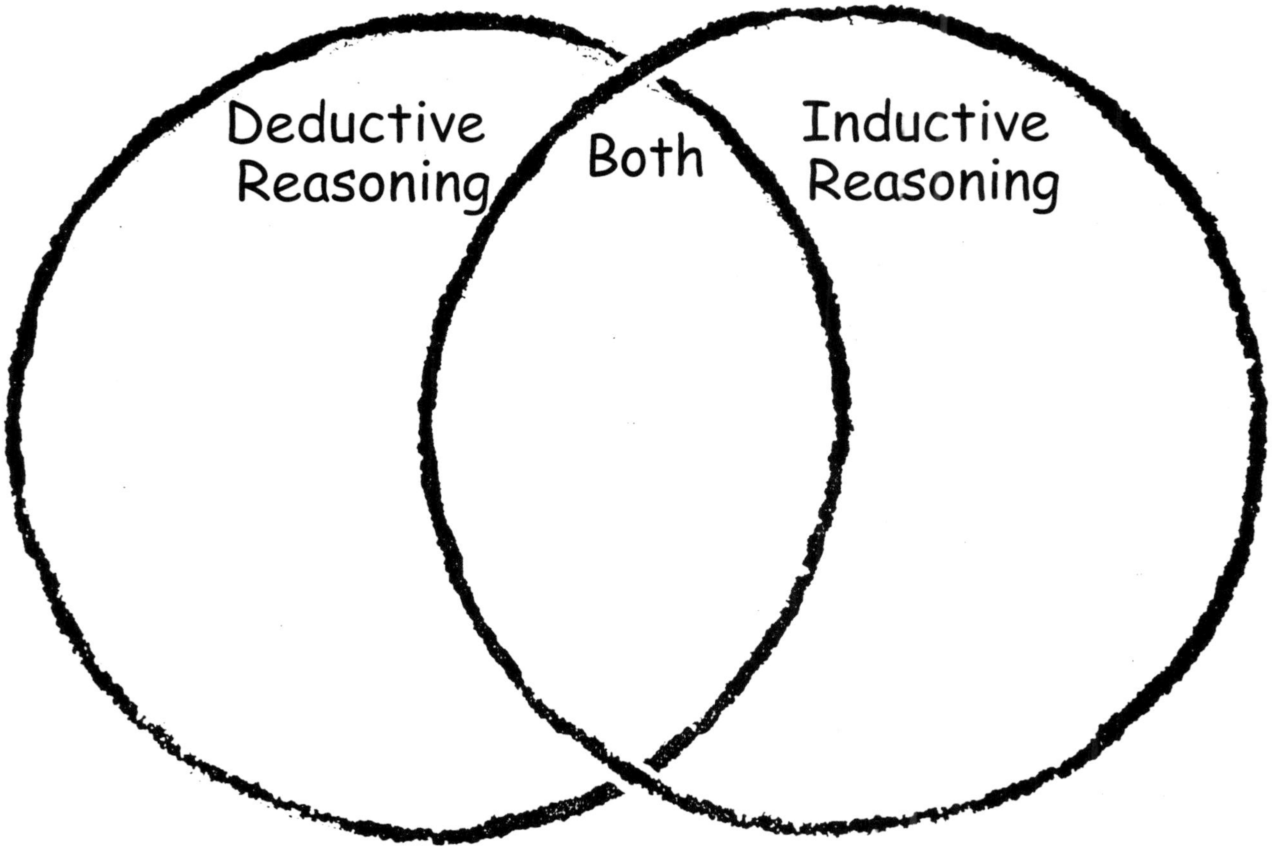

Taking Another Step: On the back of this sheet, write about a time when you used deductive or inductive reasoning in your everyday life.

Name ______________________________________ Date ____________ **Student Activity Sheet**

RP-2: Mathematics Reasoning

Introduction: **Reasoning** is drawing conclusions from facts. Reasoning is fundamental to mathematics. Whatever problem you are tackling, the reasoning process is basically the same. In this activity, you will focus on the steps in mathematics reasoning.

Directions: Complete the left-hand side of the diagram first. Then apply the steps to a math problem, and detail how you used each step in the right-hand side of the diagram.

Steps in Mathematics Reasoning

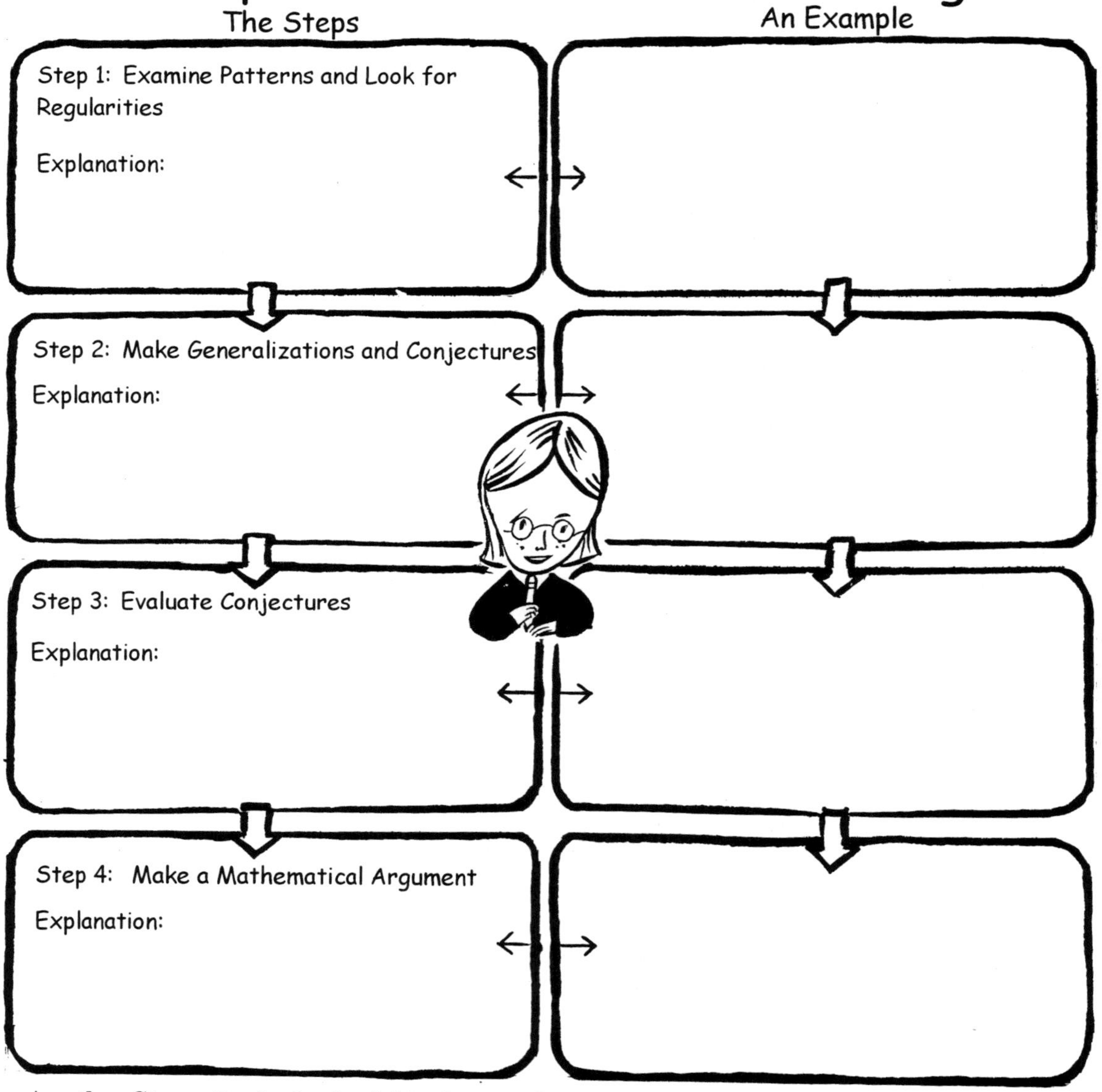

Taking Another Step: On the back of this sheet, write a paragraph explaining this statement: "Reasoning is fundamental to mathematics."

VIII. Communication

CM-1: Mathematics Communication Skills

Teaching Notes

Objective: Students will identify and discuss the use of communication skills in a mathematical context.

Key Questions:

- What are good mathematics communication skills?
- Why is each one important?
- How are these skills related to life outside of math class?

Usage Notes:

- This graphic organizer works well as a whole-class activity.
- Cite, or have students cite, classroom examples of each skill being put to use.

CM-2: A Letter Explaining a Mathematics Concept

Objective: Students will write a letter explaining a mathematics concept, and then evaluate how well they communicated.

Key Questions:

- How difficult do you think this assignment will be?
- What made it easy or difficult for you?
- How well did you communicate?
- How does expressing a mathematics concept in writing help you understand and remember it?
- How does expressing a mathematics concept in writing help you understand the use of symbols in mathematics?

Usage Notes: This graphic organizer works well as a homework activity. Assign individual students different concept topics. Consider assigning concepts that particular students are having difficulty with.

CM-3: Working in a Group

Objective: Students will identify and apply 7 group-work skills.

Key Questions:

- What group-work skills did you identify?
- Why is each skill important?
- How can you be sure to apply each skill?
- What should you do if someone else in the group is not using these skills?

Usage Notes: Consider assigning this graphic organizer in conjunction with a specific group project.

Name ____ Date ______________ **Student Activity Sheet**

CM-1: Mathematics Communication Skills

Introduction: If you think about it, you'll realize that communication is a big part of math. Your teacher communicates to your class, to groups, and to you individually, both through speaking and through grading. You communicate with your teacher by asking questions, giving answers, and turning in your assignments. You communicate with your classmates when you cooperate to solve problems. You even communicate with yourself!

Directions: Think about what makes for good communication in your math class. Then complete the diagram by identifying and describing seven good communication skills you should put to work in your math class.

Good Communication in Math Class

Taking Another Step: Review your diagram. Place a star (*) by the communication skills you do well. For the others, write one thing you can do to improve each one.

Name ______________________________ Date __________ **Student Activity Sheet**

CM-2: A Letter Explaining a Mathematics Concept

Introduction: Your math teacher has a tough job. He or she has to explain what can be difficult and complicated concepts. How well would you do it? You might never be a math teacher, but playing the part is a good way to see how much you've learned, and to help you remember it.

Directions: In the space below, write a letter to your teacher or to someone else explaining an important concept you learned in math class.

Taking Another Step: Give your letter to someone else. After she or he has read it, quiz her or him on the concept you explained. Then ask yourself: How well did I communicate this concept? What could I have done to communicate more effectively?

Name ____ Date ____________ **Student Activity Sheet**

CM-3: Working in a Group

Introduction: Working with other students is something you'll do throughout your school career, and beyond. Do you know what it takes to be an effective group member?

Directions: Each of the students below is having a *different* thought about what it takes to be good member of a work group in math class. Write the things they are thinking about.

Taking Another Step: The next time you work in a group, do your best to apply the skills you identified in the diagram. After your work with the group is finished, write a journal entry in which you evaluate how well you performed as a member of the group. Note anything you can do to improve your performance next time.

IX. Connections

CN-1: The Branches of Mathematics

Teaching Notes

Objective: Students will identify and describe the major branches of mathematics.

Key Questions:

- What are the main branches of mathematics?
- What is each branch concerned with?
- How are these branches similar? How are they different?

Usage Notes:

- This graphic organizer works well as a whole-class activity.
- Consider using this graphic organizer early in the year, as an overview, and having students refer to it periodically to keep them oriented to the area of mathematics they're working in, and to its relationship to other areas.
- Answers can vary depending upon how you divide the discipline. Answers should include arithmetic, algebra, geometry, and calculus. Other possible answers include analytical geometry/trigonometry, probablility and statistics, set theory, and logic.

CN-2: A Math Problems Constellation

Objective: Students will create a math problems constellation to see the relationships among different math problems.

Key Questions: How are the problems connected to each other?

Usage Notes:

- This graphic organizer works well as a whole-class activity.
- Be sure to model a math problems constellation on the board to ensure student understanding of the concept.

CN-3: How I Use Math in Other Subjects

Objective: Students will identify ways that math is used in other school subjects.

Key Questions:

- In what other subjects do you use, or could you use, math?
- Why is math important in each subject?

Usage Notes: Consider having students keep math-use logs in their other classes prior to completing this graphic organizer.

Name ____________________________ Date ____________ **Student Activity Sheet**

CN-1: The Branches of Mathematics

Introduction: Mathematics is divided into many different branches. The branches of math are concerned with different types of problems and how the solutions to those problems are put to use. But the branches, at their root, have more similarities than differences. They're all part of math.

Directions: Conduct research to complete the diagram. Make sure you write a description of each branch of mathematics and an explanation of its importance.

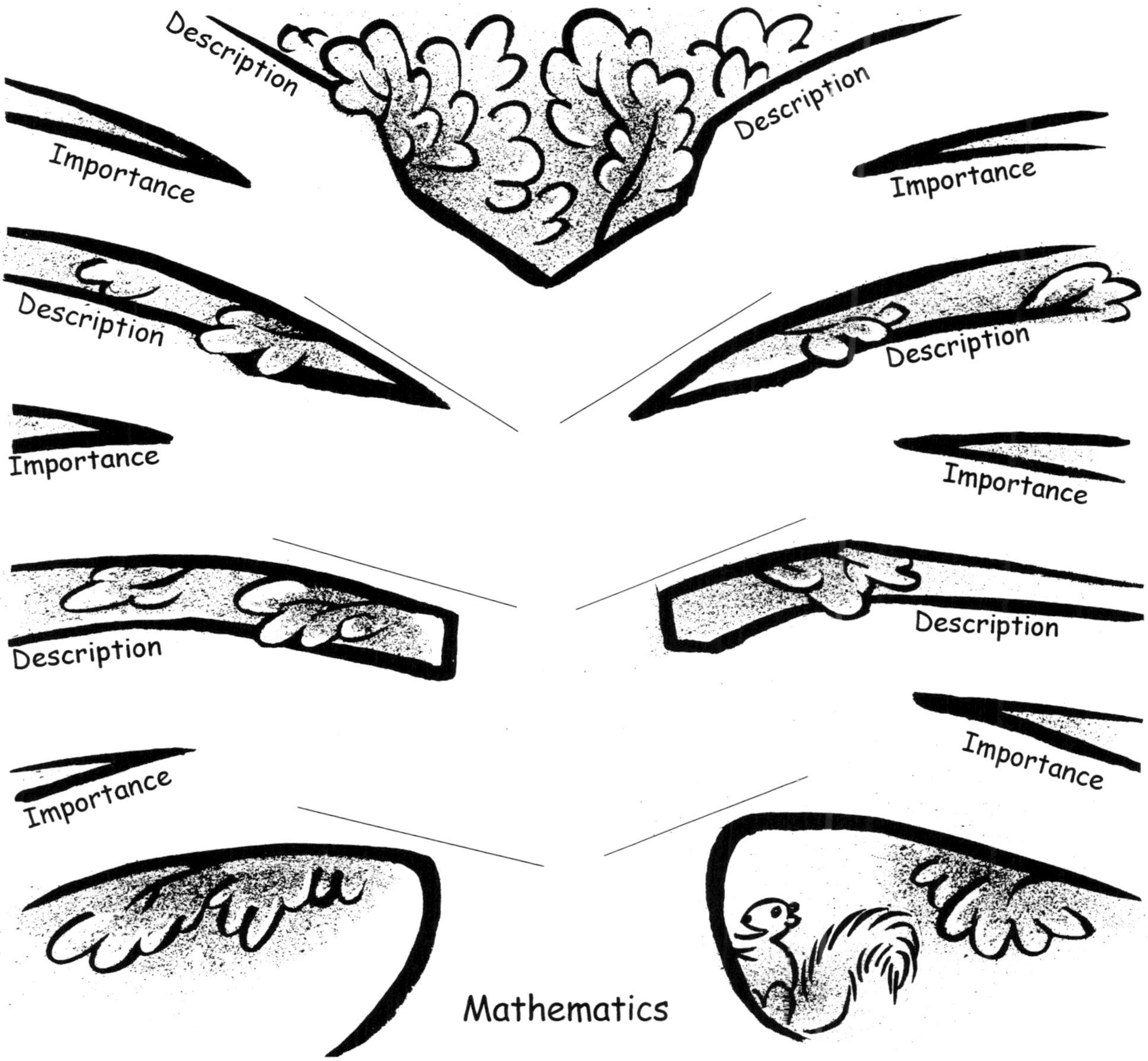

Taking Another Step: Add to your diagram by writing the names of textbook chapters, lessons, concepts, and problems on the different branches. Make sure to write them in the right places. This will show that you understand which branch of mathematics each one is a part of.

Name ________________________________ Date ____________ **Student Activity Sheet**

CN-2: A Math Problems Constellation

Introduction: A "math problems constellation" is a diagram that shows the relationships among different math problems. Each problem is represented by a circle. Lines and arrows between the circles represent the relationships between the problems. Labels on the lines define the relationships among the problems. You can expand problem constellations to include many, even hundreds, of problems. Just be sure to draw the arrows and label them with the correct relationships.

Directions: Study the example. Then, in the space below, draw a math problems constellation. Include at least three problems.

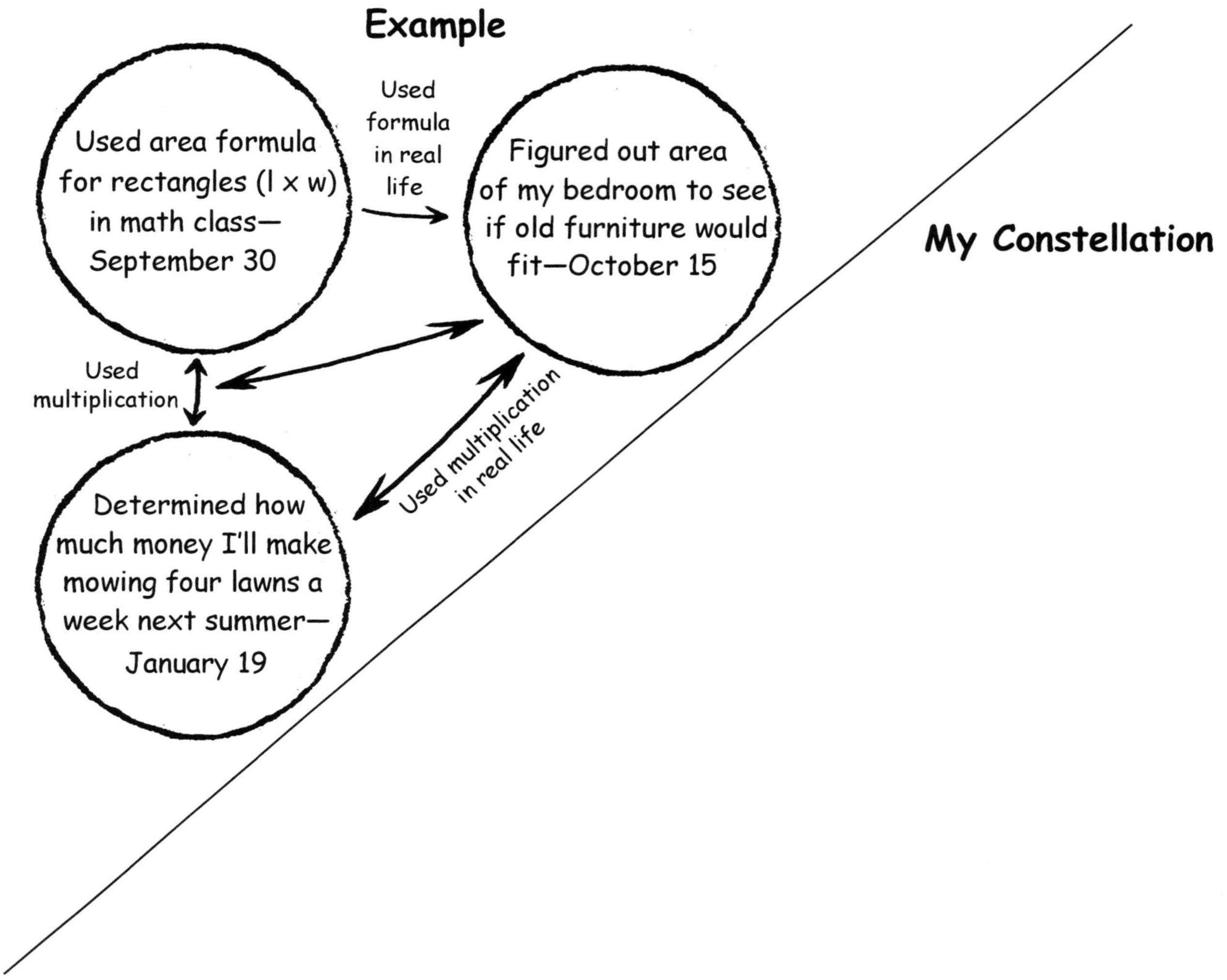

Taking Another Step: Add one problem to your constellation that seems to have absolutely nothing to do with the others. Then challenge yourself to find the connections.

Name____ Date ______________ **Student Activity Sheet**

CN-3: How I Use Math in Other Subjects

Introduction: Math isn't just for math class. You will use it in every other subject.

Directions: In each classroom, write at least two ways you use math in that subject.

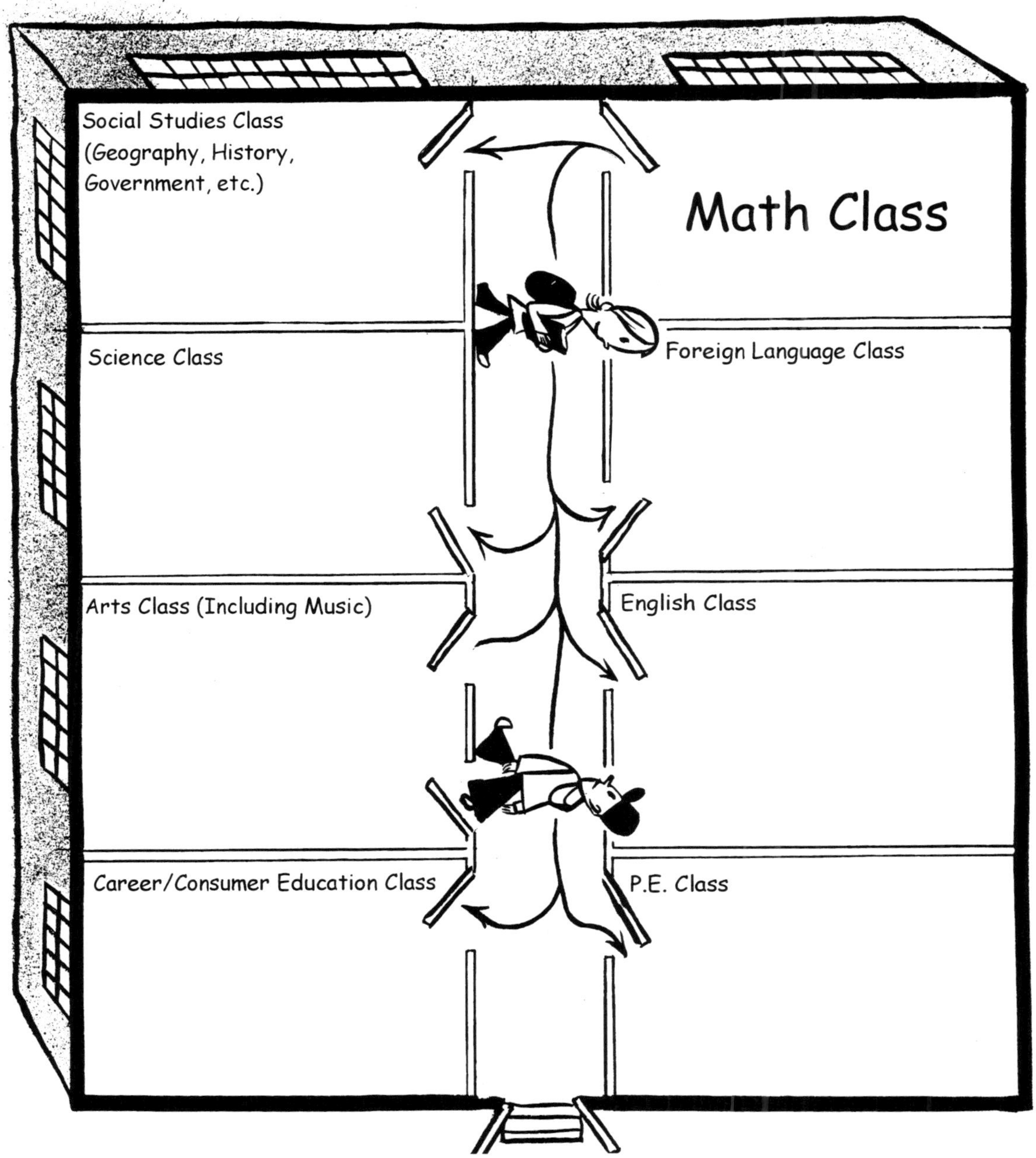

Taking Another Step: On the back of this sheet, tell about a time when you used your knowledge of mathematics to help you in another subject.

X. Representation

R-1: What Is Mathematical Representation?

Teaching Notes

Objective: Students will answer 12 fundamental questions about mathematical representation, and then ask and answer 6 more.

Key Questions: (Questions on the graphic organizer)

Usage Notes: Have students ask the questions they generate aloud, and use them as the basis for class discussions.

R-2: Types of Mathematical Representation

Objective: Students will identify several types of mathematical representation, describe them, and identify their advantages and disadvantages.

Key Questions:

- What are several types of mathematical representation?
- What are the advantages and disadvantages of each one?

Usage Notes: Work with students to develop in them a sense of *when* each type of representation is most advantageous.

Name____ Date _____________ **Student Activity Sheet**

R-1: What Is Mathematical Representation?

Introduction: **Mathematical representation** is the use of symbols, graphs, and other things to stand for mathematics ideas. You can quickly learn more about mathematical representation by asking—and answering—a few questions.

Directions: Conduct research to answer the questions in each box.

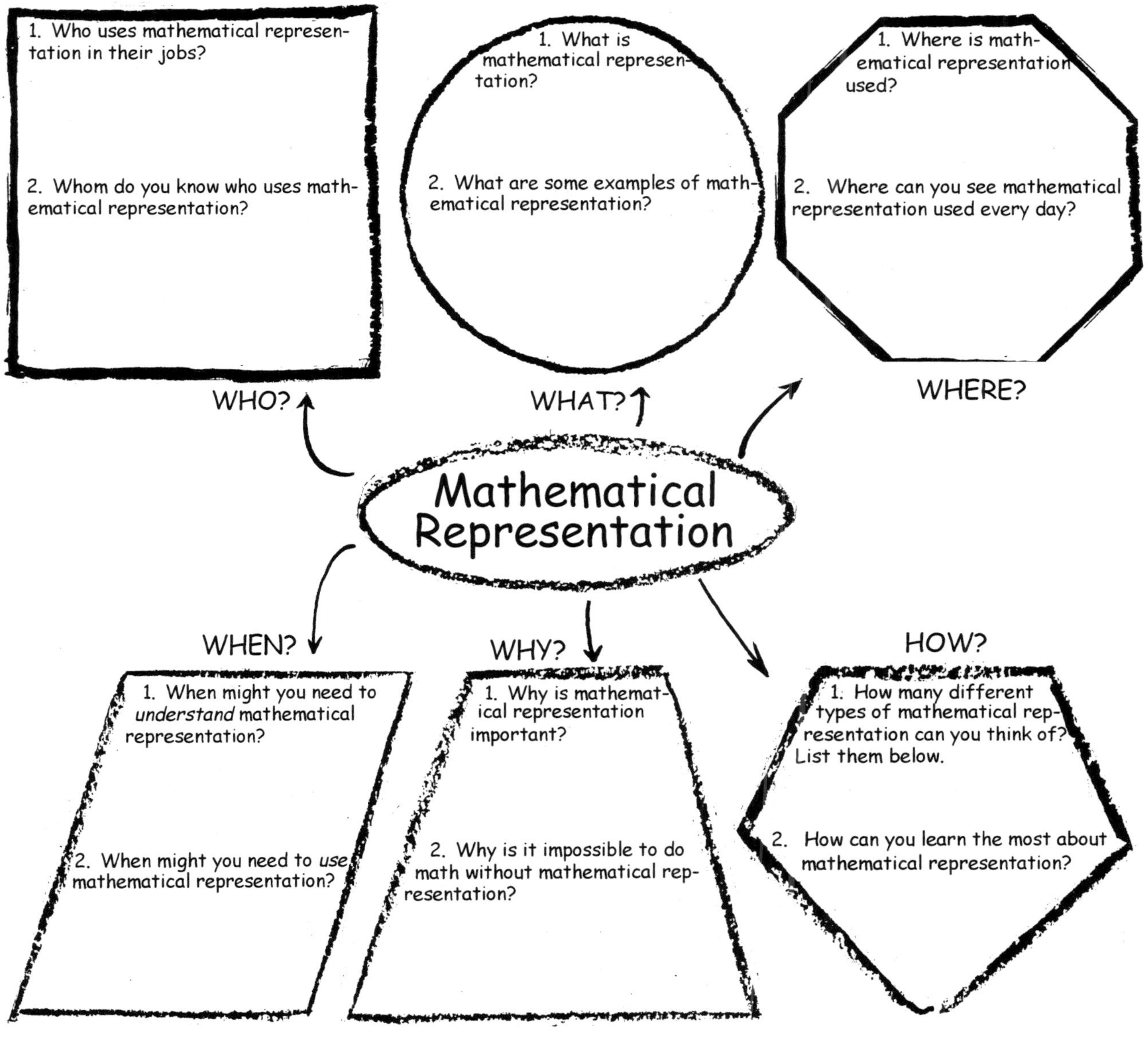

Taking Another Step: Ask a question of your own on the back of this sheet. Then trade your sheet with a partner and answer each other's questions.

Name ____ Date ____________ **Student Activity Sheet**

R-2: Types of Mathematical Representation

Introduction: There are many, many different ways to represent mathematical ideas. For example, the same idea can be represented by a formula, a table, a graph, or even by a picture. Each of them has advantages and disadvantages, strengths and weaknesses.

Directions: Complete the diagram.

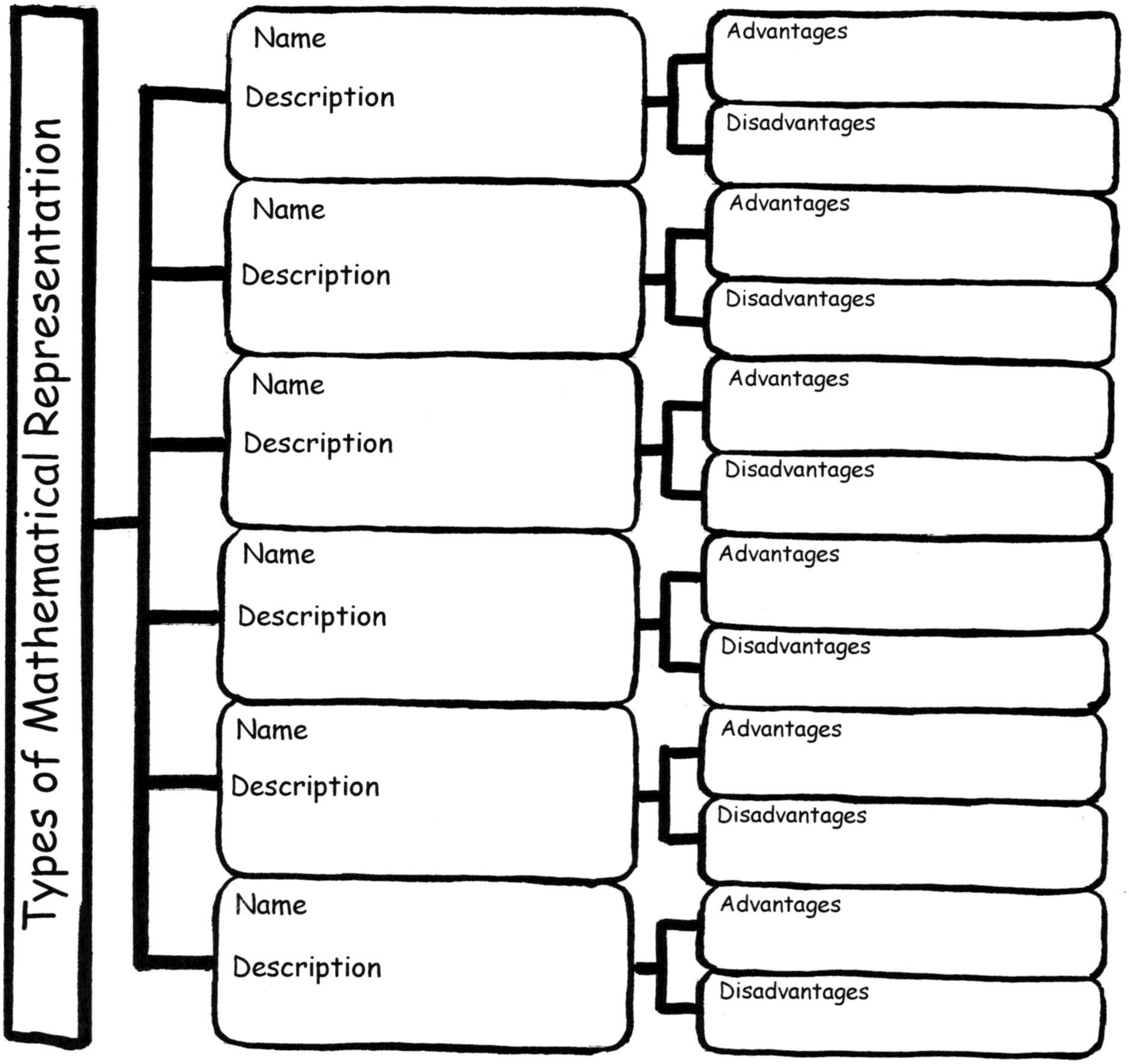

Taking Another Step: On the back of this sheet, identify a type of problem for which each type of mathematical representation would be ideal.

XI. Additional Graphic Tools

Additional Graphic Tools

Teaching Notes

The graphic organizers in this section are self-explanatory. They are also very flexible, and so you can assign them in a wide variety of classroom situations as you see fit.

Name ______________________________ Date ____________ **Student Activity Sheet**

AGT-1: Algebra Formula Log

Formula	Explanation	Application

Name ________________________________ Date ____________ **Student Activity Sheet**

AGT-2: Geometry Formula Log

Formula	Explanation	Application

Name ____ Date _____________ **Student Activity Sheet**

AGT-3: Measurement Formula Log

Formula	Explanation	Application

Name ______________________________ Date ____________ **Student Activity Sheet**

AGT-4: Math Symbol Log

Symbol	Meaning

Name____ Date ______________ **Student Activity Sheet**

AGT-5: Personal Math Glossary

Introduction: A **glossary** is a list of specialized terms and their meanings. As you progress through the school year, you will learn many new terms. This sheet gives you a place to create a personalized glossary of these terms.

Directions: As you learn new terms, record them and their meanings in the table below.

Terms Related to My Study of Mathematics	
Term	**Definition**

Name ____ Date _____________ **Student Activity Sheet**

AGT-6: Math Use Log

Introduction: You will use math in other subjects and in other areas of your life.
Directions: Keep track of how you use math outside of math class.

Using Math in Everyday Life	
Date	Description of How I Used Math Outside of Math Class

Name ______________________________ Date __________ **Student Activity Sheet**

AGT-7: The Six Questions

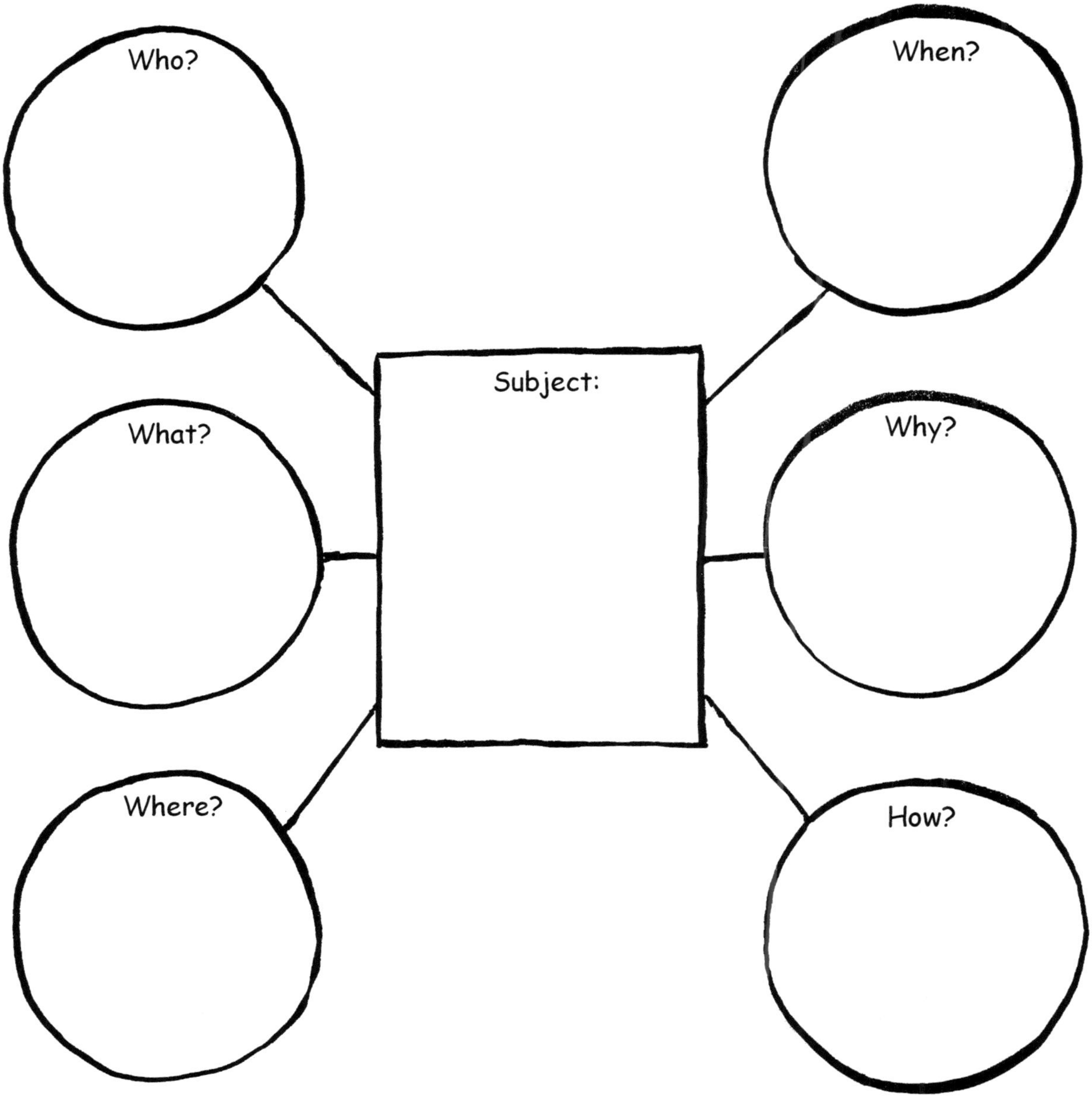

Name ____ Date _____________ **Student Activity Sheet**

AGT-8: Dealing With Math Anxiety

Introduction: You're not alone—many, many students have to deal with **math anxiety**. This term refers to the feelings of fear and frustration you sometimes get when doing your math work. By identifying these symptoms, and taking steps to eliminate them, you can lower your anxiety a great deal—and even get rid of it entirely.

Directions: List your symptoms on the left-hand side of the diagram. Then work with your teacher to come up with ways to combat them.

Symptoms of My Math Anxiety

Positive Steps I Can Take to Lessen My Anxiety

Name ______________________________ Date __________ **Student Activity Sheet**

AGT-9: Fraction Strips

$$\frac{4}{4}$$

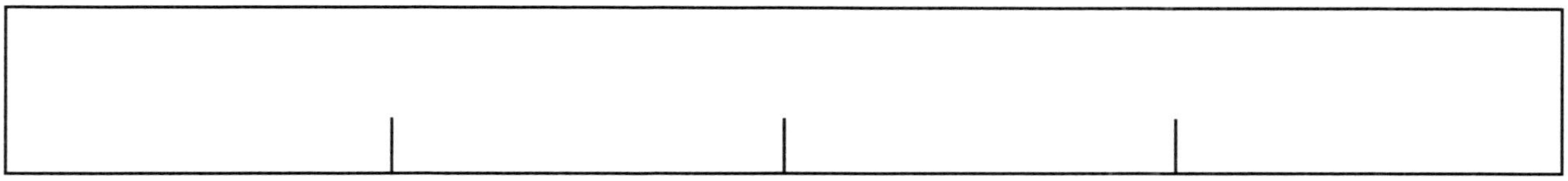

$$\frac{8}{8}$$

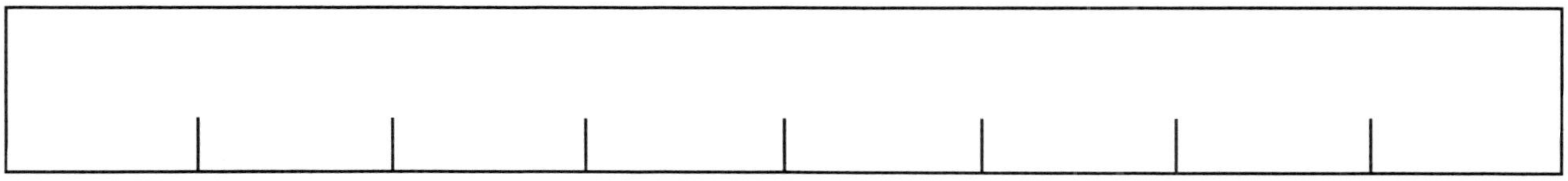

$$\frac{16}{16}$$

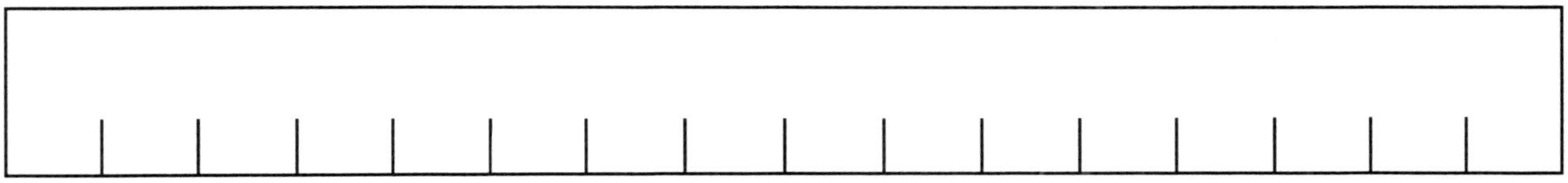

$$\frac{32}{32}$$

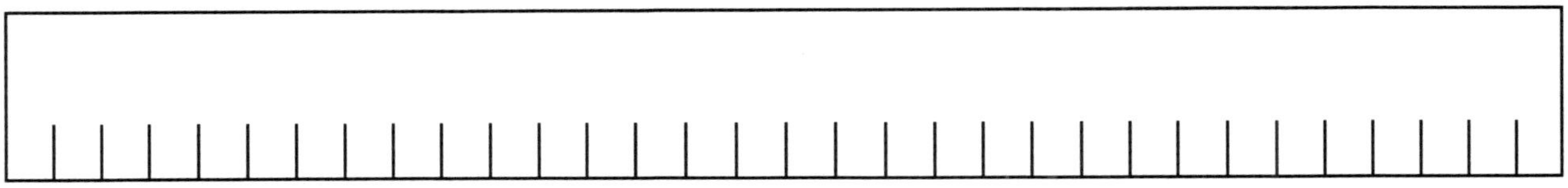

Name ______________________________ Date __________ **Student Activity Sheet**

AGT-10: Number Lines

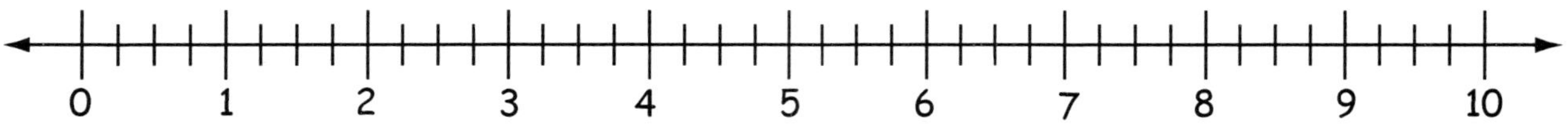

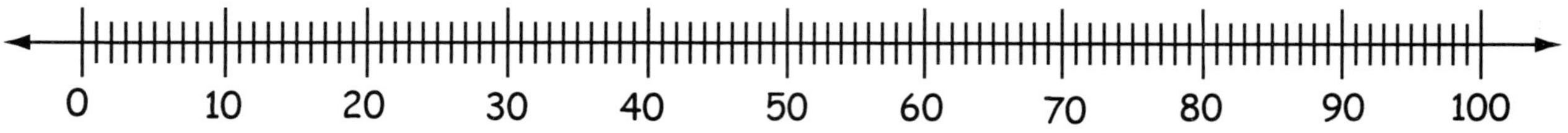

Name ______________________________________ Date ____________ **Student Activity Sheet**

AGT-11: 10 x 10 Grid

Name ______________________________ Date __________ **Student Activity Sheet**

AGT-12: 100 x 100 Grid

Name____ Date ______________

Student Activity Sheet

AGT-13: Bar Graph Template

Name____ Date____________ **Student Activity Sheet**

AGT-14: Line Graph Template

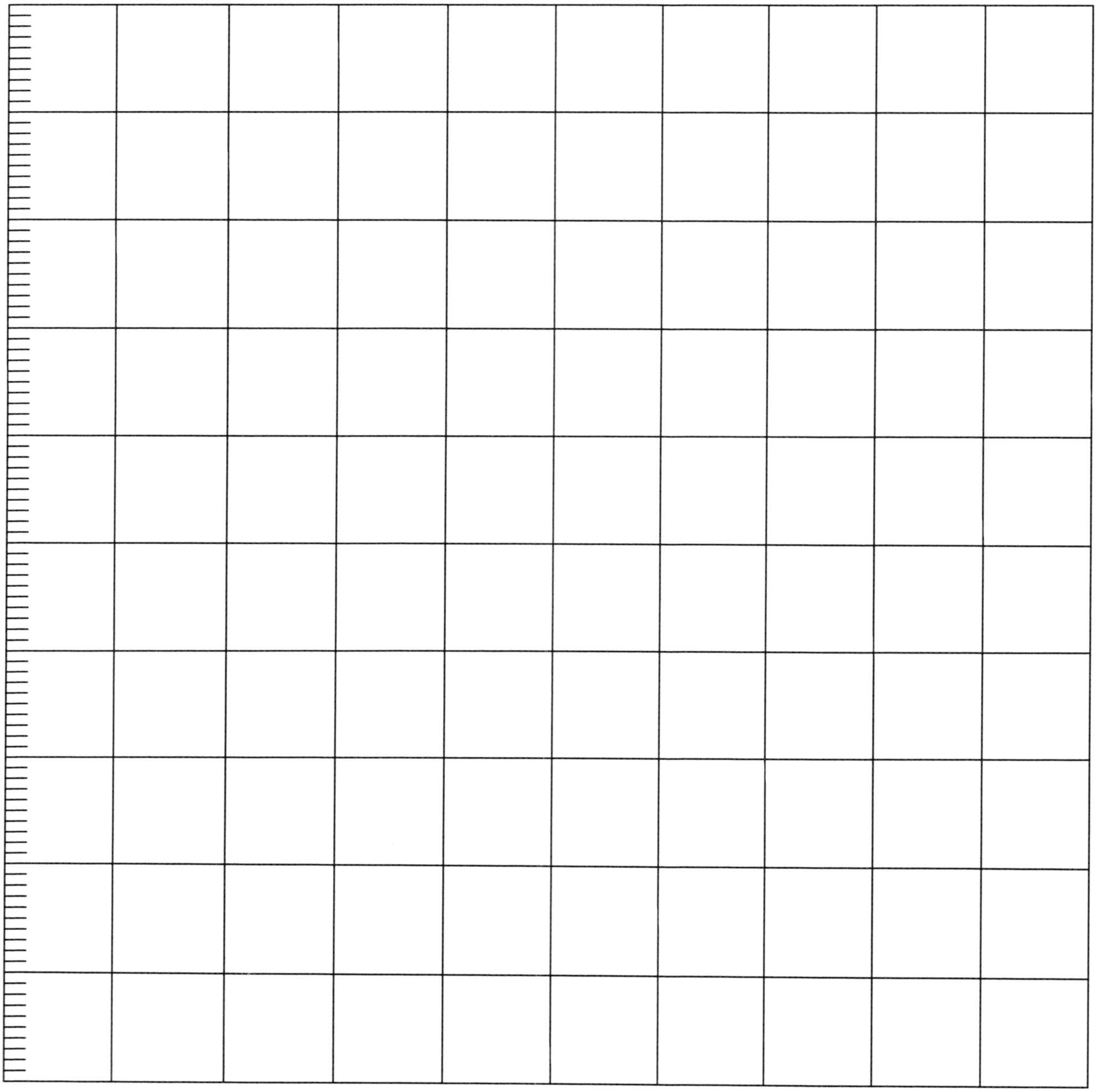

Name____ Date______________ **Student Activity Sheet**

AGT-15: Circle Graph Template

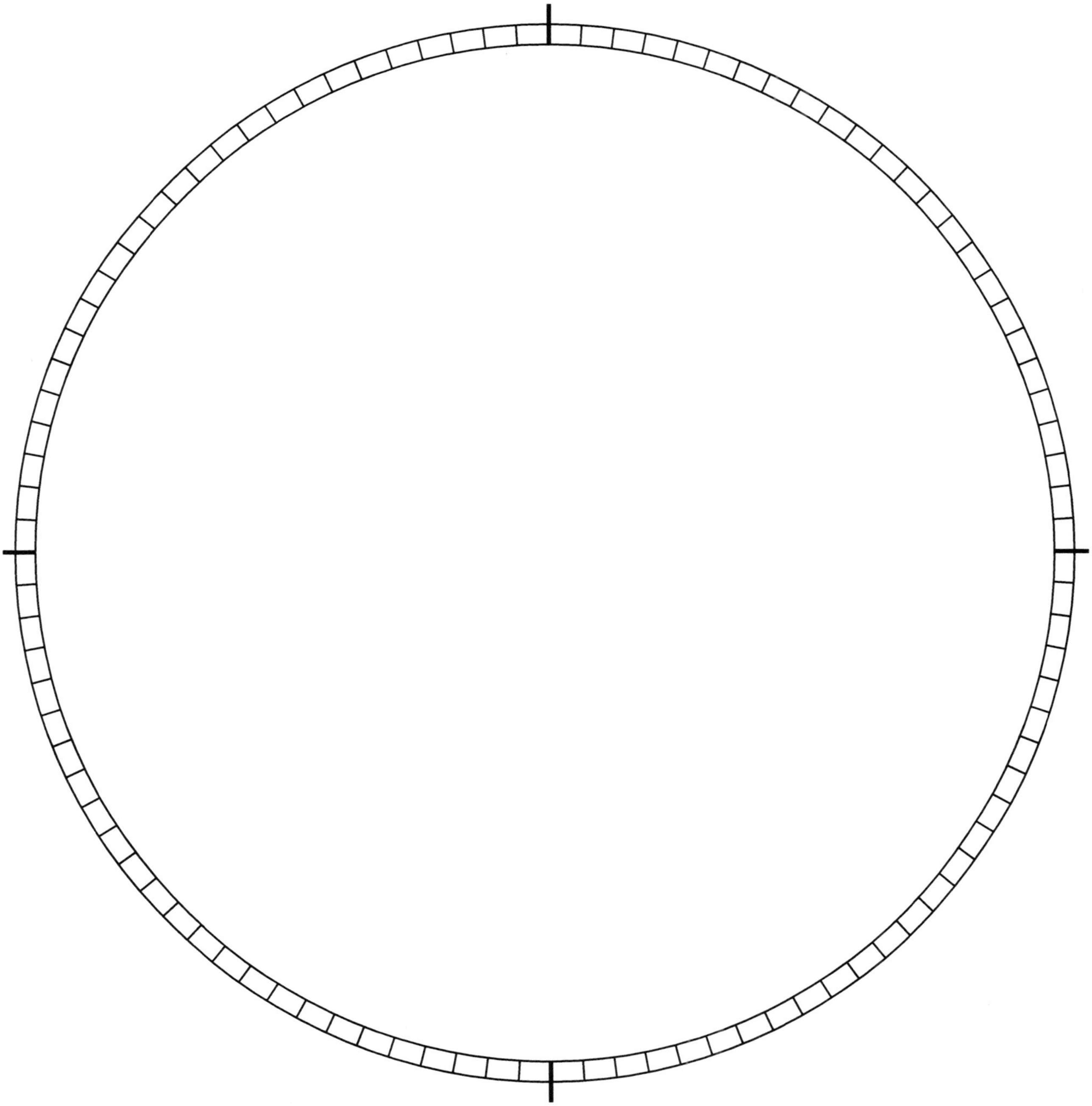

Name ____ Date ______________ **Student Activity Sheet**

AGT-16: Venn Diagram Template

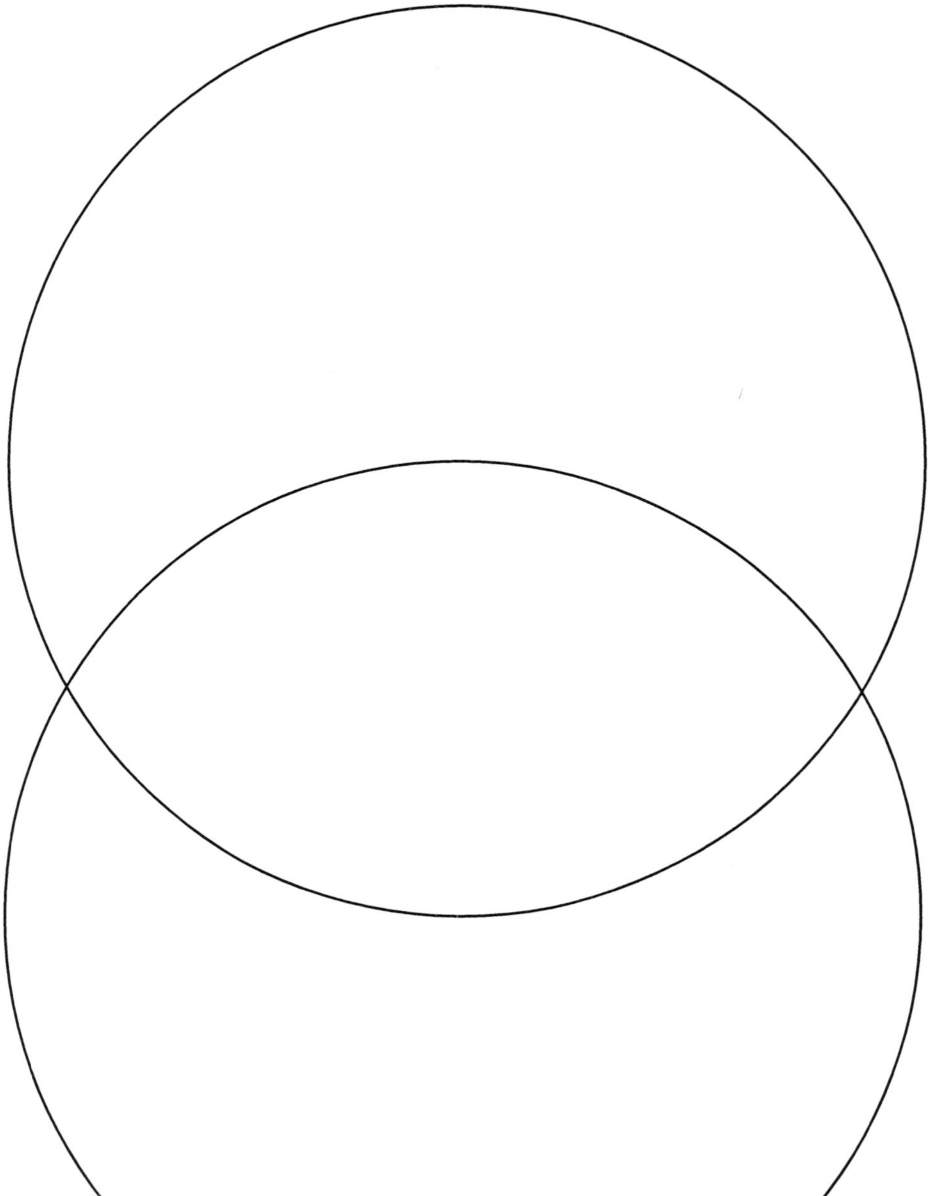

J. WESTON **WALCH** PUBLISHER

Share Your Bright Ideas with Us!

We want to hear from you! Your valuable comments and suggestions will help us meet your current and future classroom needs.

Your name____Date______________________

School name____

School address____

City ________________________State _____Zip__________Phone number (_________)______________

Grade level taught_______Subject area(s) taught____________________________Average class size_____

Where did you purchase this publication?____

Was your salesperson knowledgeable about this product? Yes_____ No_____

What monies were used to purchase this product?

___School supplemental budget ___Federal/state funding ___Personal

Please "grade" this Walch publication according to the following criteria:

Quality of service you received when purchasing	A	B	C	D	F
Ease of use	A	B	C	D	F
Quality of content	A	B	C	D	F
Page layout	A	B	C	D	F
Organization of material	A	B	C	D	F
Suitability for grade level	A	B	C	D	F
Instructional value	A	B	C	D	F

COMMENTS:____

What specific supplemental materials would help you meet your current—or future—instructional needs?

Have you used other Walch publications? If so, which ones?_________________________________

May we use your comments in upcoming communications? ___Yes ___No

Please **FAX** this completed form to **207-772-3105**, or mail it to:

Product Development, J. Weston Walch, Publisher, P. O. Box 658, Portland, ME 04104-0658

We will send you a **FREE GIFT** as our way of thanking you for your feedback. **THANK YOU!**